Pitman Research Notes in Mathematics Series

Main Editors
H. Brezis, Université de Paris
R. G. Douglas, State University of New York at Stony Brook
A. Jeffrey, University of Newcastle-upon-Tyne *(Founding Editor)*

Editorial Board
R. Aris, University of Minnesota
A. Bensoussan, INRIA, France
W. Bürger, Universität Karlsruhe
J. Douglas Jr, University of Chicago
R. J. Elliott, University of Hull
G. Fichera, Università di Roma
R. P. Gilbert, University of Delaware
R. Glowinski, Université de Paris
K. P. Hadeler, Universität Tübingen
K. Kirchgässner, Universität Stuttgart
B. Lawson, State University of New York at Stony Brook
W. F. Lucas, Cornell University
R. E. Meyer, University of Wisconsin-Madison
J. Nitsche, Universität Freiburg
L. E. Payne, Cornell University
G. F. Roach, University of Strathclyde
J. H. Seinfeld, California Institute of Technology
I. N. Stewart, University of Warwick
S. J. Taylor, University of Virginia

Submission of proposals for consideration
Suggestions for publication, in the form of outlines and representative samples, are invited by the Editorial Board for assessment. Intending authors should approach one of the main editors or another member of the Editorial Board, citing the relevant AMS subject classifications. Alternatively, outlines may be sent directly to the publisher's offices. Refereeing is by members of the board and other mathematical authorities in the topic concerned, throughout the world.

Preparation of accepted manuscripts
On acceptance of a proposal, the publisher will supply full instructions for the preparation of manuscripts in a form suitable for direct photo-lithographic reproduction. Specially printed grid sheets are provided and a contribution is offered by the publisher towards the cost of typing. Word processor output, subject to the publisher's approval, is also acceptable.

Illustrations should be prepared by the authors, ready for direct reproduction without further improvement. The use of hand-drawn symbols should be avoided wherever possible, in order to maintain maximum clarity of the text.

The publisher will be pleased to give any guidance necessary during the preparation of a typescript, and will be happy to answer any queries.

Important note
In order to avoid later retyping, intending authors are strongly urged not to begin final preparation of a typescript before receiving the publisher's guidelines and special paper. In this way it is hoped to preserve the uniform appearance of the series.

Longman Scientific & Technical
Longman House
Burnt Mill
Harlow, Essex, UK
(tel (0279) 26721)

Titles in this series

1 Improperly posed boundary value problems
A Carasso and A P Stone

2 Lie algebras generated by finite dimensional ideals
I N Stewart

3 Bifurcation problems in nonlinear elasticity
R W Dickey

4 Partial differential equations in the complex domain
D L Colton

5 Quasilinear hyperbolic systems and waves
A Jeffrey

6 Solution of boundary value problems by the method of integral operators
D L Colton

7 Taylor expansions and catastrophes
T Poston and I N Stewart

8 Function theoretic methods in differential equations
R P Gilbert and R J Weinacht

9 Differential topology with a view to applications
D R J Chillingworth

10 Characteristic classes of foliations
H V Pittie

11 Stochastic integration and generalized martingales
A U Kussmaul

12 Zeta-functions: An introduction to algebraic geometry
A D Thomas

13 Explicit *a priori* inequalities with applications to boundary value problems
V G Sigillito

14 Nonlinear diffusion
W E Fitzgibbon III and H F Walker

15 Unsolved problems concerning lattice points
J Hammer

16 Edge-colourings of graphs
S Fiorini and R J Wilson

17 Nonlinear analysis and mechanics: Heriot-Watt Symposium Volume I
R J Knops

18 Actions of fine abelian groups
C Kosniowski

19 Closed graph theorems and webbed spaces
M De Wilde

20 Singular perturbation techniques applied to integro-differential equations
H Grabmüller

21 Retarded functional differential equations: A global point of view
S E A Mohammed

22 Multiparameter spectral theory in Hilbert space
B D Sleeman

24 Mathematical modelling techniques
R Aris

25 Singular points of smooth mappings
C G Gibson

26 Nonlinear evolution equations solvable by the spectral transform
F Calogero

27 Nonlinear analysis and mechanics: Heriot-Watt Symposium Volume II
R J Knops

28 Constructive functional analysis
D S Bridges

29 Elongational flows: Aspects of the behavi
of model elasticoviscous fluids
C J S Petrie

30 Nonlinear analysis and mechanics: Heriot-Watt Symposium Volume III
R J Knops

31 Fractional calculus and integral transforms
generalized functions
A C McBride

32 Complex manifold techniques in theoretic
physics
D E Lerner and P D Sommers

33 Hilbert's third problem: scissors congruen
C-H Sah

34 Graph theory and combinatorics
R J Wilson

35 The Tricomi equation with applications to
theory of plane transonic flow
A R Manwell

36 Abstract differential equations
S D Zaidman

37 Advances in twistor theory
L P Hughston and R S Ward

38 Operator theory and functional analysis
I Erdelyi

39 Nonlinear analysis and mechanics: Heriot-Watt Symposium Volume IV
R J Knops

40 Singular systems of differential equations
S L Campbell

41 N-dimensional crystallography
R L E Schwarzenberger

42 Nonlinear partial differential equations in
physical problems
D Graffi

43 Shifts and periodicity for right invertible operators
D Przeworska-Rolewicz

44 Rings with chain conditions
A W Chatters and C R Hajarnavis

45 Moduli, deformations and classifications
compact complex manifolds
D Sundararaman

46 Nonlinear problems of analysis in geometry and mechanics
M Atteia, D Bancel and I Gumowski

47 Algorithmic methods in optimal control
W A Gruver and E Sachs

48 Abstract Cauchy problems and functional differential equations
F Kappel and W Schappacher

49 Sequence spaces
W H Ruckle

50 Recent contributions to nonlinear partial differential equations
H Berestycki and H Brezis

51 Subnormal operators
J B Conway

52 Wave propagation in viscoelastic media
F Mainardi

53 Nonlinear partial differential equations a
their applications: Collège de France Seminar. Volume I
H Brezis and J L Lions

54 Geometry of Coxeter groups
H Hiller
55 Cusps of Gauss mappings
T Banchoff, T Gaffney and C McCrory
56 An approach to algebraic K-theory
A J Berrick
57 Convex analysis and optimization
J-P Aubin and R B Vintner
58 Convex analysis with applications in the differentiation of convex functions
J R Giles
59 Weak and variational methods for moving boundary problems
C M Elliott and J R Ockendon
60 Nonlinear partial differential equations and their applications: Collège de France Seminar. Volume II
H Brezis and J L Lions
61 Singular systems of differential equations II
S L Campbell
62 Rates of convergence in the central limit theorem
Peter Hall
63 Solution of differential equations by means of one-parameter groups
J M Hill
64 Hankel operators on Hilbert space
S C Power
65 Schrödinger-type operators with continuous spectra
M S P Eastham and H Kalf
66 Recent applications of generalized inverses
S L Campbell
67 Riesz and Fredholm theory in Banach algebra
B A Barnes, G J Murphy, M R F Smyth and T T West
68 Evolution equations and their applications
F Kappel and W Schappacher
69 Generalized solutions of Hamilton-Jacobi equations
P L Lions
70 Nonlinear partial differential equations and their applications: Collège de France Seminar. Volume III
H Brezis and J L Lions
71 Spectral theory and wave operators for the Schrödinger equation
A M Berthier
72 Approximation of Hilbert space operators I
D A Herrero
73 Vector valued Nevanlinna Theory
H J W Ziegler
74 Instability, nonexistence and weighted energy methods in fluid dynamics and related theories
B Straughan
75 Local bifurcation and symmetry
A Vanderbauwhede
76 Clifford analysis
F Brackx, R Delanghe and F Sommen
77 Nonlinear equivalence, reduction of PDEs to ODEs and fast convergent numerical methods
E E Rosinger
78 Free boundary problems, theory and applications. Volume I
A Fasano and M Primicerio
79 Free boundary problems, theory and applications. Volume II
A Fasano and M Primicerio
80 Symplectic geometry
A Crumeyrolle and J Grifone
81 An algorithmic analysis of a communication model with retransmission of flawed messages
D M Lucantoni
82 Geometric games and their applications
W H Ruckle
83 Additive groups of rings
S Feigelstock
84 Nonlinear partial differential equations and their applications: Collège de France Seminar. Volume IV
H Brezis and J L Lions
85 Multiplicative functionals on topological algebras
T Husain
86 Hamilton-Jacobi equations in Hilbert spaces
V Barbu and G Da Prato
87 Harmonic maps with symmetry, harmonic morphisms and deformations of metrics
P Baird
88 Similarity solutions of nonlinear partial differential equations
L Dresner
89 Contributions to nonlinear partial differential equations
C Bardos, A Damlamian, J I Díaz and J Hernández
90 Banach and Hilbert spaces of vector-valued functions
J Burbea and P Masani
91 Control and observation of neutral systems
D Salamon
92 Banach bundles, Banach modules and automorphisms of C^*-algebras
M J Dupré and R M Gillette
93 Nonlinear partial differential equations and their applications: Collège de France Seminar. Volume V
H Brezis and J L Lions
94 Computer algebra in applied mathematics: an introduction to MACSYMA
R H Rand
95 Advances in nonlinear waves. Volume I
L Debnath
96 FC-groups
M J Tomkinson
97 Topics in relaxation and ellipsoidal methods
M Akgül
98 Analogue of the group algebra for topological semigroups
H Dzinotyiweyi
99 Stochastic functional differential equations
S E A Mohammed
100 Optimal control of variational inequalities
V Barbu
101 Partial differential equations and dynamical systems
W E Fitzgibbon III
102 Approximation of Hilbert space operators. Volume II
C Apostol, L A Fialkow, D A Herrero and D Voiculescu
103 Nondiscrete induction and iterative processes
V Ptak and F-A Potra

104 Analytic functions – growth aspects
O P Juneja and G P Kapoor
105 Theory of Tikhonov regularization for Fredholm equations of the first kind
C W Groetsch
106 Nonlinear partial differential equations and free boundaries. Volume I
J I Díaz
107 Tight and taut immersions of manifolds
T E Cecil and P J Ryan
108 A layering method for viscous, incompressible L_p flows occupying R^n
A Douglis and E B Fabes
109 Nonlinear partial differential equations and their applications: Collège de France Seminar. Volume VI
H Brezis and J L Lions
110 Finite generalized quadrangles
S E Payne and J A Thas
111 Advances in nonlinear waves. Volume II
L Debnath
112 Topics in several complex variables
E Ramírez de Arellano and D Sundararaman
113 Differential equations, flow invariance and applications
N H Pavel
114 Geometrical combinatorics
F C Holroyd and R J Wilson
115 Generators of strongly continuous semigroups
J A van Casteren
116 Growth of algebras and Gelfand–Kirillov dimension
G R Krause and T H Lenagan
117 Theory of bases and cones
P K Kamthan and M Gupta
118 Linear groups and permutations
A R Camina and E A Whelan
119 General Wiener–Hopf factorization methods
F-O Speck
120 Free boundary problems: applications and theory, Volume III
A Bossavit, A Damlamian and M Fremond
121 Free boundary problems: applications and theory, Volume IV
A Bossavit, A Damlamian and M Fremond
122 Nonlinear partial differential equations and their applications: Collège de France Seminar. Volume VII
H Brezis and J L Lions
123 Geometric methods in operator algebras
H Araki and E G Effros
124 Infinite dimensional analysis–stochastic processes
S Albeverio
125 Ennio de Giorgi Colloquium
P Krée
126 Almost-periodic functions in abstract spaces
S Zaidman
127 Nonlinear variational problems
A Marino, L Modica, S Spagnolo and M Degiovanni
128 Second-order systems of partial differential equations in the plane
L K Hua, W Lin and C-Q Wu
129 Asymptotics of high-order ordinary differential equations
R B Paris and A D Wood
130 Stochastic differential equations
R Wu
131 Differential geometry
L A Cordero
132 Nonlinear differential equations
J K Hale and P Martinez-Amores
133 Approximation theory and applications
S P Singh
134 Near-rings and their links with groups
J D P Meldrum
135 Estimating eigenvalues with *a posteriori/a pr* inequalities
J R Kuttler and V G Sigillito
136 Regular semigroups as extensions
F J Pastijn and M Petrich
137 Representations of rank one Lie groups
D H Collingwood
138 Fractional calculus
G F Roach and A C McBride
139 Hamilton's principle in continuum mechanics
A Bedford
140 Numerical analysis
D F Griffiths and G A Watson
141 Semigroups, theory and applications. Volur
H Brezis, M G Crandall and F Kappel
142 Distribution theorems of L-functions
D Joyner
143 Recent developments in structured continu
D De Kee and P Kaloni
144 Functional analysis and two-point differenti operators
J Locker
145 Numerical methods for partial differential equations
S I Hariharan and T H Moulden
146 Completely bounded maps and dilations
V I Paulsen
147 Harmonic analysis on the Heisenberg nilpo Lie group
W Schempp
148 Contributions to modern calculus of variati
L Cesari
149 Nonlinear parabolic equations: qualitative properties of solutions
L Boccardo and A Tesei
150 From local times to global geometry, contr and physics
K D Elworthy
151 A stochastic maximum principle for optima control of diffusions
U G Haussmann

U G Haussmann

The University of British Columbia

A stochastic maximum principle for optimal control of diffusions

Longman Scientific & Technical

Copublished in the United States with
John Wiley & Sons, Inc., New York

Longman Scientific & Technical
Longman Group UK Limited
Longman House, Burnt Mill, Harlow
Essex CM20 2JE, England
and Associated Companies throughout the world.

Copublished in the United States with
John Wiley & Sons, Inc., 605 Third Avenue, New York, NY 10158

First published 1986

AMS Subject Classifications: (main) 49–02
(subsidiary) 49B60

ISSN 0269-3674

British Library Cataloguing in Publication Data
Haussmann, U. G.
A stochastic maximum principle for optimal
control of diffusions.—(Pitman research
notes in mathematics, ISSN 0269–3674; 151)
1. Markov processes 2. Diffusion processes
—Mathematical models 3. Control theory
4. Stochastic analysis
I. Title
519.2′33 QA274.75

ISBN 0-582-98893-4

Library of Congress Cataloging-in-Publication Data
Haussmann, U. G.
A stochastic maximum principle for optimal control
of diffusions.

(Pitman research notes in mathematics, ISSN 0269-3674;
151)
Bibliography: p.
1. Control theory. 2. Mathematical optimization.
3. Stochastic processes. I. Title. II. Series.
QA402.3.H36 1986 629.8′312 86-19074
ISBN 0-470-20786-8 (USA only)

Printed and bound in Great Britain by
Biddles Ltd, Guildford and King's Lynn

Contents

Preface

My interest in a stochastic maximum principle dates back to 1974 and I believe that a reasonably complete theory now exists. In these notes I attempt to draw together the various results scattered in the literature in addition to adding some new material, and to present the material in a manner which is useable in the classroom at an advanced level.

I wish to thank the secretarial staff of the Mathematics Department at U.B.C. for their competent typing of the manuscript, as well as my children Matthew and Jessica for enduring a father frequently engrossed in mathematics. To these two this work is dedicated.

Vancouver, B.C., July 1986 Ulrich G. Haussmann

Introduction

In these notes we shall derive necessary conditions satisfied by a solution of the following control problem: the state x satisfies the equation

$$dx_t = f(t,x_t,u_t)dt + \sigma(t,x_t)dw_t$$

with x_0 distributed according to a given law μ. Here $\{w_t\}$ is a Brownian motion. The control u_t is of the feedback form $u(t,x)$, with values in a given set U, and is chosen from among all such controls to minimize

$$J_0(u) = E\{\int_0^T \ell_0(t,x_t,u_t)dt + c_0(x_T)\}$$

subject to the (soft or average) constraints

$$J_i(u) = E\{\int_0^T \ell_i(t,x_t,u_t)dt + c_i(x_T)\}$$

$$\begin{cases} = 0 & i = -m_1, \cdots, -1 \\ \leq 0 & i = 1, \cdots, m_2. \end{cases}$$

We shall specify the problem precisely in section three. The results constitute a maximum principle of the Pontryagin type for stochastic differential equations. The approach is to use strong perturbations of the optimal control u^*, active over a short time interval, with the corresponding perturbed state defined via a Girsanov transformation of the drift. A Lagrange multiplier theorem is used to handle the constraints. The resultant necessary condition contains an adjoint process which is defined via a martingale representation theorem. We can identify this adjoint process explicitly in the case of complete information, and can then treat some examples. Finally we show that in the linear-convex case these necessary conditions are sufficient.

A novel feature of the approach is that the admissible controls are of the feedback type, so in terms of the deterministic problem we have already solved the synthesis problem! This fortunate coincidence is completely dependent on the stochastic nature of the problem, and requires some non-degeneracy of $\sigma(t,x)$. It also complicates the proof substantially. On the other hand it is not totally surprising, since a similar non-degeneracy hypothesis guarantees that the Hamilton Jacobi-Bellman equation has a smooth solution so that a feedback control exists, cf. Fleming and Rishel (1975). In the deterministic case nothing of the sort is true.

In case σ is degenerate, there are alternatives available, Kushner (1965,1972), Bismut (1973), Arkin and Saksonov (1979), Bensoussan (1982, 1982a), but they all assume that $u_t = u(t,\omega)$, a stochastic process adapted to a fixed filtration, normally that which is generated by $\{w_t\}$, i.e. $u_t = v(t,w)$ for some function v. From the viewpoint of an engineer, such controls are not satisfactory since it is impossible to construct a functional of the unknown $\{w_t\}$, so that these are not the controls relative to which an optimal one is sought. We investigate this question further in section 12.

This material was first presented as an advanced graduate level course at the Université Pierre et Marie Curie, Paris, and subsequently at the University of British Columbia, Vancouver. The notes are reasonably self-contained and assume as a prerequisite a basic course in stochastic processes up to the Itô integral. Although no knowledge of martingale theory is required, we do need four results from this theory: Doob's inequality, the Burkholder-Davis-Gundy inequality, the Kunita-Watanabe decomposition, and the fact that Itô integrals can be time-changed into Brownian motion. These results are presented without proofs in the appendices. At the end of each section we make some historical comments and we give some exercises. The latter range from the trivial to the difficult, but the reader will find them instructive.

Since each of the twelve chapters is rather short we refer to them as sections. There are two numbering systems used: **8.2** refers to the second subsection in section eight (a corollary), whereas **(8.2)** refers to the second numbered equation of section 8.

I would like to thank Mr. P. Loewen for his careful reading of these notes and for his many suggestions for improvements. And I would especially like to thank Mme. N. El Karoui for arranging my visit to Paris, without which these notes would not exist.

0 Background

We wish to present here the basic terminology and facts about stochastic processes which we shall use. $(\Omega,\underline{\underline{F}},P)$ will denote a probability space. Let I be a closed interval in $\mathbf{R}$. If $\{\underline{\underline{F}}_t\}_{t\in I}$ is a family of σ-algebras such that for $s \leq t$, s,t in I, $\underline{\underline{F}}_s \subset \underline{\underline{F}}_t \subset \underline{\underline{F}}$ then $\{\underline{\underline{F}}_t\}$ is a filtration and $(\Omega,\underline{\underline{F}},\{\underline{\underline{F}}_t\},P)$ is a probability space with filtration, or a filtered probability space. Note that frequently we drop the index set I and write $\{\underline{\underline{F}}_t\}$. If M is a complete separable metric space (i.e. a Polish space, usually $\mathbf{R}^n$), we let $\underline{\underline{B}}(M)$ be the σ-algebra of Borel subsets of M. A (measurable) stochastic process is a mapping

$$x: I \times \Omega \to M$$

which is measurable $\underline{\underline{B}}(I) \times \underline{\underline{F}} \to \underline{\underline{B}}(M)$. We say that for fixed ω, the mapping $t \to x(t,\omega)$ is a sample path or trajectory. Moreover we usually suppress the argument ω and write t as a subscript, i.e. $x(t,\omega)$ becomes x_t. We say that the process is continuous if there is a null set N such that $t \to x_t$ is continuous for ω not in N. The process is $\{\underline{\underline{F}}_t\}$ - adapted if for each t in I, $\omega \to x(t,\omega)$ is $\underline{\underline{F}}_t$ measurable. For t in I let $\underline{\underline{F}}_t^x$ denote the smallest σ-algebra relative to which x_s is measurable for all $s \leq t$. It is called the σ-algebra generated by x up to time t and should be thought of as being the "history" of x. It follows that $\{x_t\}$ is $\{\underline{\underline{F}}_t^x\}$ - adapted.

Let τ be a random variable with values in I. We say that τ is a stopping time if for each t in I the set $\{\omega:\tau\leq t\}$ is in $\underline{\underline{F}}_t$. If τ is constant then certainly τ is a stopping time. If B is in $\underline{\underline{B}}(M)$ then the first exit time from B is

$$\tau_B = \inf\{t \geq 0: x_t \text{ not in } B\}.$$

We would like such first exit times to be stopping times for B open when x is continuous. Since the sample paths (or trajectories) of x are only continuous with probability one, we must make allowance for this in the filtration. Let $\bar{\underline{\underline{F}}}_t$ be the smallest σ-algebra containing $\underline{\underline{F}}_t$ as well as all subsets of P-negligible sets in $\underline{\underline{F}}$. Now we have

0.1 Proposition. If x is continuous, B open, then τ_B is a stopping time with respect to $\{\underline{\underline{F}}_t\}$. The proof is left as an exercise, cf. exercise 0.3.1.

We denote expectation by $E\{\cdot\}$ and conditional expectation by $E\{\cdot|\cdot\}$. The adapted process $\{x_t: t \text{ in } I\}$ is a martingale if $E|x_t| < \infty$ for t in I and if for $s \leq t$, s, t in I

$$E\{x_t|\underline{\underline{F}}_s\} = x_s \quad \text{a.s.}$$

Note that a.s. stands for almost surely and means that the event so characterized takes place with probability one. It is a supermartingale if

$$E\{x_t|\underline{\underline{F}}_s\} \leq x_s \quad \text{a.s.}$$

Finally the martingale $\{x_t\}$ is a square integrable martingale if

$$\sup_t E\,|x_t|^2 < \infty.$$

Let us now recall some facts about Brownian motion. A standard Brownian motion in $\mathbf{R}^n$ is a continuous stochastic process $\{w_t: t \geq 0\}$ with values in $\mathbf{R}^n$ such that

(i) $w_0 = 0$ a.s.

(ii) w has independent increments, i.e. for $t < s \leq r < u$, $(w_s - w_t)$ and $(w_u - w_r)$ are independent,

(iii) the increment $w_s - w_t$ has normal distribution, mean zero and covariance $I|s-t|$, where I is the $n\times n$ identity matrix.

It follows that the law of w_t has density

$$p(t,x) = (2\pi t)^{-n/2} \exp[-|x|^2/(2t)].$$

Note that here $I = [0,\infty)$, $M = \mathbf{R}^n$. Actually we shall use a slightly different definition. A standard Brownian motion on $(\Omega,\underline{\underline{F}},\{\underline{\underline{F}}_t\},P)$ with values in $\mathbf{R}^n$ is a continuous $\mathbf{R}^n$-valued process, $\{w_t: t \geq 0\}$, which satisfies (i) and (iii) as well as

ii)' w is adapted and $(w_t - w_s)$ is independent of $\underline{\underline{F}}_s$ for $t > s$.

We observe that any such process satisfies (ii) and conversely if a continuous process $\{w_t: t \geq 0\}$ satisfies (i),(ii),(iii), then it is a standard Brownian motion with $\underline{\underline{F}}_t = \underline{\underline{F}}_t^w$. We add that a standard Brownian motion in $\mathbf{R}^n$ is just an n-tuple of independent standard Brownian motions in $\mathbf{R}$.

The proof of the following proposition is an exercise in integration in the complex plane and is left to the reader, cf. exercise 0.3.2.

0.2 Proposition. A continuous $\mathbf{R}^n$-valued adapted process $\{w_t: t \geq 0\}$ satisfying (i) above is a Brownian motion on $(\Omega, \underline{\underline{F}}, \{\underline{\underline{F}}_t\}, P)$ if and only if for every λ in $\mathbf{R}^n$, $0 \leq s < t$,

$$E\{\exp[i\lambda\cdot(w_t - w_s)] | \underline{\underline{F}}_s\} = \exp[-(t-s)|\lambda|^2/2] \quad \text{a.s.}$$

We assume that the reader is familiar with the concept of a stochastic integral of the form

$$I(t) = \int_{t_0}^t b_s \cdot dw_s$$

when w is an $\mathbf{R}^n$-valued Brownian motion on $(\Omega, \underline{\underline{F}}, \{\underline{\underline{F}}_t\}, P)$. We recall that I(t) is defined if b_s is an $\mathbf{R}^n$-valued adapted process such that

$$P\{\int_{t_0}^t |b_s|^2 ds < \infty\} = 1.$$

If b_s satisfies the stronger condition

$$E \int_{t_0}^T |b_s|^2 \, ds < \infty$$

then I(t) is a continuous martingale on $[t_0, T]$, and

$$E\, I(t) = 0, \quad E\, I(t)^2 = \int_{t_0}^t E|b_s|^2 ds.$$

This fact will be used repeatedly in the sequel. Moreover if τ is a stopping time such that $t_0 \leq \tau \leq t$ a.s., and if

$$1_\tau(s) = \begin{cases} 1 & \text{if } s < \tau \\ 0 & \text{otherwise} \end{cases}$$

then

$$I(\tau) = \int_{t_0}^t 1_\tau(s) b_s \cdot dw_s.$$

Even more holds. Let $\underline{\underline{F}}_{t+} = \bigcap_{s>t} \underline{\underline{F}}_s$. If τ is a stopping time with respect to $\{\bar{\underline{\underline{F}}}_{t+}\}$, then we still have

$$I(\tau) = \int_{t_0}^{t} 1_\tau(s)\, b_s \cdot dw_s .$$

This follows since $\underline{\underline{F}}_t \subset \bar{\underline{\underline{F}}}_{t+}$, and since $\{w_t\}$ is still a Brownian motion with respect to $\{\bar{\underline{\underline{F}}}_{t+}\}$ according to exercise 0.3.5.

We conclude with a useful inequality. For all N,c

$$\text{(0.1)} \qquad P\{I(t) \geq c\} \leq Nc^{-2} + P\{\int_{t_0}^{t} |b_s|^2 ds > N\}.$$

0.3 Exercises

0.3.1 Prove proposition 0.1.

0.3.2 Prove proposition 0.2. The "only if" part is an integration in the complex plane. The "if" part uses the uniqueness of the Fourier transform.

0.3.3 Prove that a standard Brownian motion on $(\Omega, \underline{\underline{F}}, \{\underline{\underline{F}}_t\}, P)$ is a martingale.

0.3.4 Show that a supermartingale with constant mean is a martingale.

0.3.5 Prove that if $\{w_t\}$ is an $(\Omega, \underline{\underline{F}}, \{\underline{\underline{F}}_t\}, P)$ Brownian motion then it is an $(\Omega, \underline{\underline{F}}, \{\bar{\underline{\underline{F}}}_{t+}\}, P)$ Brownian motion.

0.4 Comments. The basic references here are the books by Liptser and Shiryayev (1972), Ikeda and Watanabe (1981), Dellacherie and Meyer (1975,1980). A concise but nevertheless more complete presentation of the basic results is given by Bensoussan (1982).

1 Girsanov's theorem

Let us now, for the moment, consider the control problems which we shall be studying. Generally they consist of minimizing the expected value of a functional of the "state" and the "control" processes, in the presence of constraints. The state process $\{x_t\}$ will generally be a continuous stochastic process with values in $\mathbf{R}^n$, solution of the stochastic differential equation

$$dx_t = f(t,x_t,u_t)dt + \sigma(t,x_t,u_t)dw_t$$

$$\text{i.e.} \quad x_t = x_0 + \int_0^t f(s,x_s,u_s)ds + \int_0^t \sigma(s,x_s,u_s)dw_s$$

where $\{w_t\}$ is a Brownian motion, and $\{u_t\}$ is the control process. We may assume the Itô hypotheses, i.e.

$$|f(t,x,u) - f(t,y,u)| + |\sigma(t,x,u) - \sigma(t,y,u)| \leq K|x-y|$$

$$|f(t,x,u)| + |\sigma(t,x,u)| \leq K(1+|x|)$$

If now $\{u_t\}$ is an adapted stochastic process, then the usual Itô existence theorem proves existence and uniqueness of adapted solutions of the above stochastic differential equation. But if we take a control of the form (in the scalar case, say)

$$u_t = u(x_t) = \begin{cases} 1 \text{ if } x_t > 0 \\ -1 \text{ if } x_t < 0 \end{cases}$$

then Itô's method fails, and yet we would like to define a solution x corresponding to such a control. In this section and the next we shall define a weaker notion of solution using a theorem due to Girsanov.

We are given a filtered probability space carrying an $\mathbf{R}^n$-valued Brownian motion $(\Omega, \underline{\underline{F}}, \{\underline{\underline{F}}_t\}_{0\leq t\leq T}, P, \{w_t\}_{0\leq t\leq T})$, $T < \infty$ fixed. We are also given an $\mathbf{R}^n$-valued adapted process $\{\theta_t: 0\leq t\leq T\}$ such that

(1.1) $$\int_0^T |\theta_t|^2 dt < \infty \qquad \text{a.s.}$$

Define

(1.2) $$\zeta_t^s(\theta) = \exp\{\int_s^t \theta_r \cdot dw_r - 1/2\int_s^t |\theta_r|^2 dr\}, \ s \leq t \leq T,$$

and write $\zeta_t(\theta)$ if $s = 0$. From Itô's lemma and exercise 1.5.1 it follows that $\zeta_t^s(\theta)$ is the unique $\{\underline{\underline{F}}_t\}$-adapted solution of

(1.3) $$d\zeta_r = \zeta_r\theta_r\cdot dw_r, \quad \zeta_s = 1.$$

As remarked in the last section $\{\zeta_t^s(\theta): s \leq t \leq T\}$ will be a martingale if $\int_s^T E|\zeta_t^s\theta_t|^2 dt < \infty$. A condition guaranteeing this is given in lemma 1.1. For a measurable function ϕ we write sup ϕ for the essential supremum of ϕ. Also a $\wedge$ b denotes min{a,b}.

1.1 Lemma. If $\sup_{t,\omega}|\theta_t(\omega)|^2 = c < \infty$, then ζ_t^s is an $\{\underline{\underline{F}}_t\}$ martingale on [s,T] and $E\,\zeta_t^s(\theta) = 1$.

Proof: Let $\tau_m = \inf\{t\geq s: \int_s^t \theta\cdot dw \geq m\}$. Then (1.2) implies

$$E\int_s^T|\zeta^s_{r\wedge\tau_m}\,\theta_r|^2 dr \leq \int_s^T e^{2m}c\,dr < \infty,$$

and (1.3) implies $\zeta^s_{t\wedge\tau_m}(\theta) = \zeta_t^s(1_{\tau_m}\theta)$ is a martingale. Hence

$E\,\zeta^s_{t\wedge\tau_m}(\theta) = E\,\zeta^s_{s\wedge\tau_m}(\theta) = 1$ and Fatou's lemma implies

(1.4) $$E\,\zeta_t^s(\theta) = E\lim\zeta^s_{t\wedge\tau_m} \leq \liminf E\,\zeta^s_{t\wedge\tau_m} = 1.$$

But

(1.5) $$\begin{aligned}\zeta_t^s(\theta)^2 &= \exp\{\int_s^t 2\theta\cdot dw - 1/2\int_s^t|2\theta|^2 dr\}\exp\int_s^t|\theta|^2 dr\\ &\leq \zeta_t^s(2\theta)\exp[c(t-s)].\end{aligned}$$

Now (1.4) applied to $\zeta_t^s(2\theta)$ implies

$$E\int_s^T|\zeta_t^s(\theta)\theta_t|^2 dt \leq \int_s^T e^{cT}c\,dt < \infty$$

so the result follows from (1.3). Q.E.D.

1.2 Corollary. Assume (1.1). Then $\zeta_t^s(\theta)$ is a super martingale on

$s \leq t \leq T$.

Proof: Let θ_m be a sequence of bounded, adapted process such that

$\int_0^T|\theta_m - \theta|^2 dt \to 0$ a.s. as $m \to \infty$. Fix $s \leq r \leq t \leq T$. Then

$\left(\zeta_t^s(\theta_m),\ \zeta_r^s(\theta_m)\right) \to \left(\zeta_t^s(\theta),\ \zeta_r^s(\theta)\right)$ in probability (cf. exercise 1.5.2) and by possibly taking a subsequence we may assume the convergence to be a.s. Fatou's lemma and lemma 1.1 imply

$$\begin{aligned} E\{\zeta_t^s(\theta)|\underline{\underline{F}}_r\} &= E\{\lim \zeta_t^s(\theta_m)|\underline{\underline{F}}_r\} \\ &\leq \liminf E\{\zeta_t^s(\theta_m)|\underline{\underline{F}}_r\} \\ &= \liminf \zeta_r^s(\theta_m) \\ &= \zeta_r^s(\theta) \qquad \text{a.s.} \end{aligned}$$

Q.E.D.

1.3 Corollary. Assume (1.1). If $E\,\zeta_T^s(\theta) = 1$, then $\zeta_t^s(\theta)$ is a martingale on $[s,T]$.

Proof: $E\,\zeta_s^s = 1 = E\,\zeta_T^s$, so the supermartingale ζ_t^s has constant mean, hence is a martingale. Q.E.D.

1.4 Theorem (Girsanov). Assume (1.1) and $E\,\zeta_T(\theta) = 1$. Then $\bar{P}$ given by $\frac{d\bar{P}}{dP} = \zeta_T(\theta)$ is a probability measure on $(\Omega,\underline{\underline{F}})$. Moreover

(1.6) $$\bar{w}_t = w_t - \int_0^t \theta_s\,ds$$

is a Brownian motion on $(\Omega,\underline{\underline{F}},\{\underline{\underline{F}}_t\},\bar{P})$.

Proof: The first assertion is trivial. The second will be established if we show that for any vector λ in $\mathbf{R}^n$, $0 \leq s \leq t \leq T$, we have

(1.7) $$\bar{E}\{\exp[i\lambda\cdot(\bar{w}_t - \bar{w}_s)]\,|\underline{\underline{F}}_s\} = \exp[-|\lambda|^2(t-s)/2] \quad \text{a.s.}$$

Note that $\bar{E}$ denotes expectation with respect to $\bar{P}$. The corollaries and $E\,\zeta_T(\theta) = 1$ imply that $E\{\zeta_T^t(\theta)|\underline{\underline{F}}_t\} = 1$, so $E\,\zeta_T^t(\theta) = 1$. Let ϕ be in $L^2([0,T];\mathbf{R}^n)$, the L^2 space of $\mathbf{R}^n$ valued functions defined on $[0,T]$, and set

$$\zeta_T^t(\theta+i\phi) = \exp\{\int_t^T(\theta+i\phi)\cdot dw - 1/2\int_t^T(\theta+i\phi)\cdot(\theta+i\phi)ds\}.$$

Then (1.3) still holds, i.e. $\zeta_T^t(\theta+i\phi)$ satisfies $d\zeta = \zeta(\theta+i\phi)\cdot dw$, and

(1.8) $$|\zeta_T^t(\theta+i\phi)| = \zeta_T^t(\theta)\exp\ 1/2\int_t^T|\phi|^2ds.$$

We begin by showing that $E\{\zeta_T^t(\theta+i\phi)|\underline{\underline{F}}_t\} = 1$. Let $\{\theta_n\}$ be a sequence of bounded adapted processes such that $\int_0^T|\theta_n - \theta|^2 dt \to 0$ a.s. Then also $(\theta_n+i\phi)$ converges to $(\theta+i\phi)$ in $L^2([t,T];R^n)$ a.s., and hence $\zeta_T^t(\theta_n+i\phi) \to \zeta_T^t(\theta+i\phi)$ in probability. By (1.8) and lemma 1.1

$$E|\zeta_T^t(\theta_n+i\phi)| = \exp 1/2 \int_t^T|\phi|^2 ds,$$

and by (1.8) and the hypothesis of the theorem

$$E|\zeta_T^t(\theta+i\phi)| = \exp 1/2 \int_t^T|\phi|^2 ds = E|\zeta_T^t(\theta_n+i\phi)|. \tag{1.9}$$

Moreover from (1.8) and (1.5) and lemma 1.1

$$E|\zeta_T^t(\theta_n+i\phi)|^2 = E\ \zeta_T^t(\theta_n)^2 \exp\int_t^T|\phi|^2 ds \leq E\ \zeta_T^t(2\theta_n)\exp\int_t^T(|\theta_n|^2+|\phi|^2)dr < \infty$$

so that (1.3) implies that $\zeta_s^t(\theta_n+i\phi)$ is a martingale on $[t,T]$. Unfortunately we cannot conclude from this, as was done in corollary 1.2, that $\zeta(\theta+i\phi)$ is a martingale because Fatou's lemma no longer applies. Nevertheless the convergence in probability implies that for any $\varepsilon > 0$ there exists $\delta_n(\varepsilon) \to 0$ as $n \to \infty$ such that $P(\Omega_{n\varepsilon}) > 1 - \delta_n(\varepsilon)$ where

$$\Omega_{n\varepsilon} = \{\omega: |\zeta_T^t(\theta_n+i\phi) - \zeta_T^t(\theta+i\phi)| < \varepsilon\}. \tag{1.10}$$

Since $|\zeta_T^t(\theta+i\phi)|$ is integrable then by absolute continuity

$$\int_{\Omega-\Omega_{n\varepsilon}} |\zeta_T^t(\theta+i\phi)|dP < \varepsilon \tag{1.11}$$

if n is sufficiently large. Now using (1.9) - (1.11) we obtain for n sufficiently large

$$\begin{aligned} \int_{\Omega-\Omega_{n\varepsilon}} |\zeta_T^t(\theta_n+i\phi)|dP &= E|\zeta_T^t(\theta_n+i\phi)| - \int_{\Omega_{n\varepsilon}} |\zeta_T^t(\theta_n+i\phi)|dP \\ &\leq E|\zeta_T^t(\theta+i\phi)| - \int_{\Omega_{n\varepsilon}} [|\zeta_T^t(\theta+i\phi)| - \varepsilon]dP \\ &\leq \int_{\Omega-\Omega_{n\varepsilon}} |\zeta_T^t(\theta+i\phi)|dP + \varepsilon < 2\varepsilon. \end{aligned} \tag{1.12}$$

Hence by (1.10) - (1.12)

$$\begin{aligned} &\int_\Omega|\zeta_T^t(\theta_n+i\phi) - \zeta_T^t(\theta+i\phi)|dP \\ &\quad \leq \varepsilon + \int_{\Omega-\Omega_{n\varepsilon}} |\zeta_T^t(\theta_n+i\phi) - \zeta_T^t(\theta+i\phi)|dP < 4\varepsilon, \end{aligned}$$

i.e. $\zeta_T^t(\theta_n+i\phi) \to \zeta_T^t(\theta+i\phi)$ in L^1. But now, since

$$E\{\zeta_T^t(\theta_n+i\phi)|\underline{\underline{F}}_t\} = \zeta_t^t(\theta_n+i\phi),$$

then $E\{\zeta_T^t(\theta+i\phi)|\underline{\underline{F}}_t\} = 1$.

Next let $I(\phi) = \int_0^T \phi\cdot dw - \int_0^T \phi\cdot\theta dt$. Use exercise 1.5.3 to obtain

$$\bar{E}\{\exp\, iI(\phi)|\underline{\underline{F}}_t\} = E\{\zeta_T(\theta)\exp\, iI(\phi)|\underline{\underline{F}}_t\}/E\{\zeta_T(\theta)|\underline{\underline{F}}_t\}$$

$$= E\{\zeta_T(\theta+i\phi)|\underline{\underline{F}}_t\}\exp\{-1/2\int_0^T|\phi|^2 ds\}/\zeta_t(\theta)$$

(1.13) $$= E\{\zeta_T^t(\theta+i\phi)|\underline{\underline{F}}_t\}\zeta_t(\theta+i\phi)\exp\{-1/2\int_0^T|\phi|^2 ds\}/\zeta_t(\theta)$$

$$= \zeta_t(\theta)\exp\{1/2\int_0^t|\phi|^2 ds + iI(1_t\phi)\}\exp\{-1/2\int_0^T|\phi|^2 ds\}/\zeta_t(\theta)$$

$$= \exp\{iI(1_t\phi)\}\exp\{-1/2\int_t^T|\phi|^2 ds\}$$

where $1_t(s) = \begin{cases} 1 & \text{if } s < t \\ 0 & \text{otherwise.}\end{cases}$

Let $\{e_1,\cdots,e_n\}$ be the standard basis for $\mathbf{R}^n$. Then

$$I(1_t e_k) = w_k(t) - \int_0^t \theta_k(s)ds = \bar{w}_k(t)$$

and (since I is linear)

$$\lambda\cdot(\bar{w}_t - \bar{w}_s) = \sum_k \lambda_k\big(\bar{w}_k(t) - \bar{w}_k(s)\big) = I\big(\lambda(1_t - 1_s)\big).$$

From (1.13) [n.b. $I(0) = 0$]

$$\bar{E}\{\exp[i\lambda\cdot(\bar{w}_t - \bar{w}_s)]|\underline{\underline{F}}_s\} = \bar{E}\{\exp[iI\big(\lambda(1_t-1_s)\big)]|\underline{\underline{F}}_s\}$$

$$= \exp\{iI\big(\lambda(1_t-1_s)1_s\big)\}\exp\big\{-1/2\int_s^T|\lambda|^2|1_t-1_s|^2 dr\big\}$$

$$= \exp\{-1/2|\lambda|^2(t-s)\}$$

and (1.7) is established. Q.E.D.

1.5 Exercises

1.5.1 Prove the uniqueness of the solution of (1.3) by computing the differential of the ratio of two solutions.

1.5.2 Prove that if θ_n are bounded, adapted processes such that

$$\lim_{n\to\infty} \int_s^T|\theta_n - \theta|^2 dt = 0 \qquad \text{a.s.}$$

with θ adapted, satisfying (1.1) then $\zeta_t^s(\theta_n) \to \zeta_t^s(\theta)$ in probability. (0.1) may be useful.

1.5.3 Use the definition of conditional expectation to show that

$$\bar{E}\{\psi|\underline{\underline{F}}_t\}E\{\zeta_T(\theta)|\underline{\underline{F}}_t\} = E\{\psi\zeta_T(\theta)|\underline{\underline{F}}_t\}$$

for any bounded $\underline{\underline{F}}$-measurable function ψ, where $d\bar{P} = \zeta_T(\theta)dP$.

1.6 Comments. Our proof of Girsanov's theorem follows in part Bensoussan (1982) as well as Girsanov's original work (1960). A cleaner proof which is based on Lévy's representation of Brownian motion as a continuous martingale with quadratic variation equal to t, is given by Ikeda and Watanabe (1981).

Theorem 1.4 was first established by Girsanov (1960). Another proof was then given by Beneš (1971) when he applied the theorem to obtain existence of stochastic optimal controls. A version of the theorem with w replaced by a martingale can be found in Ikeda and Watanabe (1981).

2 Weak solutions

We shall investigate how Girsanov's theorem relates to Itô equations. We begin with some notation. $C^n \equiv C(0,T;\mathbf{R}^n)$ is the space of continuous functions $[0,T] \to \mathbf{R}^n$. Let $\underline{\underline{G}}^n = \underline{\underline{B}}\big(C(0,T;\mathbf{R}^n)\big)$ and let $\underline{\underline{G}}_t^n$ be the sub-σ-algebra generated by the family of sets

$$\big\{\{x \text{ in } C(0,T;\mathbf{R}^n): x_s \text{ in } B\}: 0 \le s \le t,\ B \text{ in } \underline{\underline{B}}(\mathbf{R}^n)\big\}.$$

Then $\{\underline{\underline{G}}_t^n\}$ is the canonical Borel filtration and $\underline{\underline{G}}_T^n = \underline{\underline{G}}^n$.

We are given a filtered probability space $(\Omega,\underline{\underline{F}},\{\underline{\underline{F}}_t\}_{0\le t\le T},P)$ carrying an $\mathbf{R}^d$-valued Brownian motion $\{w_t: 0\le t\le T\}$, as well as an $\mathbf{R}^n$-valued continuous adapted process $\{x_t: 0\le t\le T\}$ satisfying

(2.1) $$dx_t = b(t,x)dt + a(t,x_t)dw_t$$

where x_0 is an $\underline{\underline{F}}_0$-measurable random variable,

$$b: [0,T]\times C^n \to \mathbf{R}^n$$

$$a: [0,T]\times\mathbf{R}^n \to \mathbf{R}^n\otimes\mathbf{R}^d$$

and b, a are Borel measurable, with b $\{\underline{\underline{G}}_t^n\}$-adapted. For x in C^n, let

$$\|x\|_t = \sup\{|x(s)|: 0 \le s \le t\}.$$

We begin with a lemma.

2.1 <u>Lemma</u>. Assume there exist constants k_0, k_1 such that

(2.2) $$\int_0^T |a(t,x)|^2 dt \le k_0 \qquad x \text{ in } \mathbf{R}^n$$

(2.3) $$|b(t,x)|^2 \le k_1(1+\|x\|_t^2) \qquad x \text{ in } C^n.$$

If $M_t = \int_0^t a(s,x_s)dw_s$ and $p \ge 1$, then for some $K < \infty$

(2.4) $$\|x\|_t^p \le K\big(1 + |x_0|^p + \|M\|_t^p\big) \quad \text{a.s.}$$

Proof: Observe first that (2.2) implies that $\{M_t\}$ is defined and is a square-integrable martingale. In fact (2.2) is more than adequate, but below we need this stronger hypothesis so we prefer to state it at once. From (2.1) there exists a constant $\bar{K}$ such that

$$\|x\|_t^2 \le 3\{|x_0|^2 + T\int_0^t |b|^2 ds + \|M\|_t^2\}$$
$$\le \bar{K}\{1 + |x_0|^2 + \int_0^t \|x\|_s^2 ds + \|M\|_t^2\}.$$

By Gronwall's inequality (cf. exercise 2.7.1)

$$\|x\|_t^2 \le \bar{K}(1+|x_0|^2+\|M\|_t^2) + \bar{K}\int_0^t e^{\bar{K}(t-s)}\bar{K}(1+|x_0|^2+\|M\|_s^2)ds$$
$$\le \bar{K}(1+|x_0|^2+\|M\|_t^2)e^{\bar{K}t}.$$

The result follows readily since $(1+a)^q \le 2^q(1+a^q)$ if $a \ge 0$, $q \ge 0$.

Q.E.D.

Now assume that $\{\theta_t\}$ is an $\mathbf{R}^n$-valued adapted process such that $\int_0^T|\theta_t|^2dt < \infty$ a.s. The following theorem gives an L^p bound on $\zeta_T(\theta)$.

2.2 Theorem. Assume (2.2), (2.3) and assume there exist $\varepsilon > 0$, ξ, K_0 finite such that for any s in [0,T]

(2.5) $$E \exp \varepsilon|x_0|^2 < \infty$$

(2.6) $$\int_0^s|\theta_t|^2dt \le \xi(1+\|x\|_s^2), \quad |\theta_s|^2 \le K_0(1+\|x\|_s^2) \quad \text{a.s.}$$

Then there exists $\xi_0 > 0$ depending on ε, k_0, K_0, k_1 and n such that if $(p^2-p)\xi < \xi_0$ then $E\zeta_T(\theta)^p \le k$ where k depends only on ε,n and p, but not on θ or ξ.

Proof: Let $\tau_N = \inf\{t\ge 0: |\int_0^t\theta\cdot dw| \ge N \text{ or } \int_0^t|\theta|^2ds \ge N\}$. For any $p > 0$, since $E\int_0^T 1_{\tau_N}(t)\zeta_t(p\theta)^2|\theta_t|^2dt \le Ne^{2pN}$, then

$$E\, \zeta_{T\wedge\tau_N}(p\theta) = 1 + E\int_0^{T\wedge\tau_N} \zeta_{t\wedge\tau_N}(p\theta)\, p\theta_t\cdot dw_t = 1.$$

From (2.6) and (2.4)

$$\zeta_{T\wedge\tau_N}(\theta)^p = \zeta_{T\wedge\tau_N}(p\theta)\exp[1/2(p^2-p)\int_0^{T\wedge\tau_N}|\theta|^2dt]$$

(2.7) $$\le \zeta_{T\wedge\tau_N}(p\theta)\exp\, 1/2(p^2-p)\xi K(1+|x_0|^2+\|M\|_{T\wedge\tau_N}^2).$$

Recall that $\zeta_{t\wedge\tau_N}(p\theta) = \zeta_t(1_{\tau_N} p\theta)$.

Let $\bar{w}_t^N = w_t - \int_0^{t\wedge\tau_N} p\theta_s ds$, so $\bar{w}_t^N$ is a Brownian motion under $d\bar{P} = \zeta_{T\wedge\tau_N}(p\theta)dP$. Let $\bar{M}_t^N = \int_0^t a(t,x_t)d\bar{w}_t^N$. Then $M_t = \bar{M}_t^N + \int_0^{t\wedge\tau_N} pa\theta ds$ and from (2.2), (2.3) and (2.4), (2.6) we obtain

$$\|M\|_t^2 \leq 2\|\bar{M}^N\|_t^2 + 2p^2\ k_0 \int_0^{t\wedge\tau_N} K_0(1+\|x\|_s^2)ds$$

$$\leq K\{\|\bar{M}^N\|_t^2 + 1 + |x_0|^2 + \int_0^t \|M\|_s^2 ds\},$$

so again by Gronwall's inequality

$$\|M\|_T^2 \leq K\left(1+|x_0|^2 + \|\bar{M}^N\|_T^2\right).$$

Note that the constant K may vary from equation to equation. Substituting this into (2.7), and taking expectation gives

$$E\zeta_{T\wedge\tau_N}(\theta)^p \leq \bar{E}\ \exp[1/2(p^2-p)\xi K(1+|x_0|^2+\|\bar{M}^N\|_T^2)]$$

$$\leq \exp\{1/2(p^2-p)\xi K\}\bar{E}\{\exp(p^2-p)\xi K|x_0|^2\}^{1/2}\bar{E}\{\exp(p^2-p)\xi K\|\bar{M}^N\|_T^2\}^{1/2}.$$

But if $a_i(t)$ is the ith row of $a(t,x_t)$, then

$$\|\bar{M}^N\|_T^2 \leq \sum_i \|\int_0^{\cdot} a_i(t)\cdot d\bar{w}_t^N\|_T^2 \leq \sum_i \|\beta_i^N\|_{R_i(T)}^2$$

where $\beta_i^N(t) = \int_0^{S_i(t)} a_i \cdot d\bar{w}^N$ and

$$S_i(t) = \inf\{s \geq 0: R_i(s) = t\}$$

with $R_i(s) = \int_0^s |a_i(t)|^2 dt$. It follows from lemma 2.6.4 that β_i^N is a Brownian motion on the random interval $[0, R_i(T)]$. Now Hölder's inequality implies

$$\bar{E}\ \exp[\lambda\|\bar{M}^N\|_T^2] \leq \bar{E}\ \prod_i \exp[\lambda\|\beta_i^N\|_{R_i(T)}^2]$$

$$\leq \prod_i \bar{E}\{\exp[n\lambda\|\beta_i^N\|_{R_i(T)}^2]\}^{1/n}$$

$$= \prod_i \bar{E}\{\exp[n\lambda\|\beta\|_{R_i(T)}^2]\}^{1/n}$$

where β is a scalar Brownian motion on $(\Omega,\underline{\underline{F}},\bar{P})$. Since $R_i(T) \leq k_0$ by (2.2), then

$$\bar{E}\ \exp[\lambda\|\bar{M}^N\|_T^2] \leq \bar{E}\ \exp[n\lambda\|\beta\|_{k_0}^2] \equiv e_0 < \infty$$

if $\lambda n k_0 < 1/2$, cf. exercise 2.7.7. If $(p^2-p)\xi K \leq \varepsilon$ and $(p^2-p)\xi K < (2nk_0)^{-1}$ then

$$E\zeta_{T\wedge\tau_N}(\theta)^p \leq e^{\varepsilon/2} E\{\exp\ \varepsilon|x_0|^2\}^{1/2} e_0^{1/2} \equiv k.$$

Fatou's lemma gives $E\zeta_T(\theta)^p \leq k$. Q.E.D.

2.3 Corollary. Assume (2.2), (2.3),(2.5) and $|\theta_t|^2 \leq K_0(1+\|x\|_t^2)$ a.s. Then there exists $p > 1$ and $k < \infty$ depending only on ε and p such that $E\zeta_T(\theta)^p \leq k$.

Proof: (2.6) is satisfied with $\xi = K_0 T$. Hence for p sufficiently close to 1 we have $(p^2-p)K_0 T < \xi_0$. The result follows from theorem 2.2.

Q.E.D.

2.4 Corollary. Assume (2.2),(2.3),(2.5) and $|\theta_t|^2 \leq K_0(1+\|x\|_t^2)$ a.s. Then $E\zeta_T(\theta) = 1$.

Proof: As in the proof of theorem 2.2 $E\zeta_{T\wedge\tau_N}(\theta) = 1$. Since $\zeta_{T\wedge\tau_N}(\theta) \to \zeta_T(\theta)$ a.s. and by corollary 2.3 is uniformly integrable cf. appendix 2.6 then the convergence is also in L^1. Q.E.D.

2.5 Remark. Under the assumptions of corollary 2.4, Girsanov's theorem is applicable, so that

(2.8) $$dx_t = [b(t,x) + a(t,x_t)\theta_t]dt + a(t,x_t)d\bar{w}_t$$

with $\bar{w}$ Brownian motion under $\bar{P}$; in other words $\{x_t\}$ is a solution of

(2.8) on $(\Omega,\underline{\underline{F}},\{\underline{\underline{F}}_t\},\bar{P},\{\bar{w}_t\})$ where $d\bar{P} = \zeta_T(\theta)dP$. It is called a <u>weak</u> solution. We say that

$$dx = \tilde{b}(t,x)dt + \tilde{a}(t,x)dw \tag{2.9}$$

has a weak solution if there exists a filtered probability space $(\Omega,\underline{\underline{F}},\{\underline{\underline{F}}_t\},P)$ carrying an adapted process $\{x_t\}$ and a standard Brownian motion $\{w_t\}$, such that (2.9) is satisfied. Then $\{x_t\}$ is said to be a weak solution of (2.9). The main point is that the space and the Brownian motion are <u>not</u> specified a priori, and that x_t is $\{\underline{\underline{F}}_t\}$ adapted but not necessarily $\{\underline{\underline{F}}_t^w\}$ adapted, i.e. it need not be a strong solution. There is a well-known counterexample due to Cirel'son, cf. Liptser and Shiryayev (1977) or Ikeda and Watanabe (1981). The solution also need not be pathwise unique, i.e. there may exist two solutions $\{x_t\}$, $\{x_t'\}$ on some $(\Omega,\underline{\underline{F}},\{\underline{\underline{F}}_t\},\bar{P},\{\bar{w}_t\})$ such that

$$\bar{P}\{\|x-x'\|_T = 0\} < 1.$$

However <u>law uniqueness</u> holds, i.e. if $\{x^i\}$ are solutions of (2.8) on $(\Omega^i,\underline{\underline{F}}^i,\{\underline{\underline{F}}_t^i\},\bar{P}^i,\{\bar{w}_t^i\})$, $i = 1,2$, then the law of $\{x_t^1\}$ on C^n is the same as the law of $\{x_t^2\}$, provided law uniqueness holds for solutions of (2.1) and $\theta_t(\omega) = \theta(t,x(\omega))$ (in addition to the hypotheses of corollary 2.4). The proof can be found in Ikeda and Watanabe (1981), page 179-180.

As we have just pointed out, Girsanov's theorem provides one way to obtain weak solutions of some equations: obtain in some fashion (say by Itô's existence theorem) a weak solution of a related equation (2.1), and change the probability measure from P to $\bar{P}$. Then the solution of (2.1) will be a solution of (2.8). If this happens to be the desired equation then we may rejoice, but note that the recipe may fail: try to solve $dx = \tilde{b}(x)dt$ where $\tilde{b}$ is discontinuous! Nevertheless if for example our state satisfies

$$dx = u\,dt + dw \tag{2.10}$$

with u as given at the beginning of section 1, then we can obtain a (weak) solution of (2.10) as follows. Let $\{w_t\}$ be a Brownian motion on some space (Ω,F,P) and let $x_t = x_0 + w_t$. Then $\{x_t\}$ is a weak solution of (2.10): take $\theta_t = u_t$.

2.6 Appendix. We recall here some elementary facts about uniform integrability. A sequence of functions $\{f_n\}$ is uniformly integrablo if

$$\lim_{N\to\infty} \sup_n \int_{\{|f_n|\geq N\}} |f_n| dP = 0.$$

Lebesgue's convergence theorem states:

2.6.1 If $f_n \to f$ in measure, and if $\{|f_n|^p\}$ is uniformly integrable, then $f_n \to f$ in L^p.

Here are two sufficient conditions for uniform integrability, the first one is commonly encountered in the statement of Lebesgue's theorem, whereas the second is used throughout this work.

2.6.2 If $|f_n| \leq g$ and g is in L^1, then $\{f_n\}$ is uniformly integrable.

2.6.3 If for some $p > 1$

$$\sup_n \int |f_n|^p dP < \infty$$

then $\{f_n\}$ is uniformly integrable.

We conclude by stating a result about stochastic integrals whose proof can be found in McKean (1969).

2.6.4 Lemma. If $\{w(t): 0 \leq t \leq T\}$ is an $\mathbf{R}^n$ valued Brownian motion and $\{b(t): 0 \leq t \leq T\}$ is an $\mathbf{R}^n$ valued stochastic process such that

$$E \int_0^T |b(t)|^2 dt < \infty$$

then $I(S(t))$ is a standard Brownian motion on $[0, R(T)]$ where

$$R(t) = \int_0^t |b(\tau)|^2 d\tau < \infty$$

$$S(t) = \inf\{\theta \geq 0: R(\theta) = t\},$$

and $I(t) = \int_0^t b(\theta)\cdot dw$.

2.7 Exercises

2.7.1 Show that if $b \geq 0$ and if for all t in $[0,T]$

$0 \leq m(t) \leq a(t) + b\int_0^t m(s)ds$

then

$m(t) \leq a(t) + b\int_0^t \exp[b(t-s)]a(s)ds.$

Hint: let $u(t) = \int_0^t m(s)ds.$

2.7.2 Let a,b be defined on $[0,T]\times\mathbf{R}^n$, Borel measurable, a bounded, invertible with bounded inverse, $x \to a(t,x)$ Lipschitz uniformly in t, and $|b(t,x)| \le K(1+|x|)$. Show that

$$dx_t = b(t,x_t)dt + a(t,x_t)dw_t$$

has a weak solution for any initial x_0 satisfying (2.5).

2.7.3 Prove 2.6.1.

2.7.4 Prove 2.6.2.

2.7.5 Prove 2.6.3.

2.7.6 Show that in the case n = 1, b(t) = c, constant, the lemma 2.6.4 reduces to the well known fact that $\{cw(t/c^2): 0 \le t \le Tc^2\}$ is a Brownian motion if $\{w(t): 0 \le t \le T\}$ is.

2.7.7 If $\{w(t): t \ge 0\}$ is a standard one dimensional Brownian motion recall that the distribution of $\|w\|_t$ has density

$$2(2\pi t)^{-1/2} \exp[-x^2/2t], \quad t \ge 0.$$

Show that $E \exp[\lambda\|w\|_t^2] < \infty$ if $2\lambda t < 1$.

2.8 Comments. There are various results concerning the validity of $E\zeta_T(\theta) = 1$. The best condition is

$$E \exp[1/2\int_0^T|\theta|^2 ds] < \infty$$

due to Novikov (1973), but the form given here is more useful for our purposes. The fact that $\zeta_T(\theta)$ is in L^p (i.e. corollary 2.3) is due to Beneš (1971) in a special case, and to Haussmann (1976) in the general case.

3 The problem

We are given a probability distribution μ on $\mathbf{R}^n$, a Borel subset U of a Euclidean space, and Borel measurable functions

$\ell_i:[0,T] \times \mathbf{R}^n \times U \to \mathbf{R}$, $c_i:\mathbf{R}^n \to \mathbf{R}$, $i = -m_1,\cdots,-1,0,1,\cdots,m_2$,

$f:[0,T] \times \mathbf{R}^n \times U \to \mathbf{R}^n$, $\sigma:[0,T] \times \mathbf{R}^n \to \mathbf{R}^n \otimes \mathbf{R}^d$.

We assume that for some $q < \infty$

$$(3.1) \qquad |\ell_i(t,x,u)| + |c_i(x)| \le k_2(1+|x|^q+|u|^q).$$

3.1 Definition. $\mathcal{U}_0$, the set of controls, consists of all measurable $\{\underline{\underline{G}}^n_t\}$-adapted $u:[0,T]\times C^n \to U$ such that there exists some probability space with filtration and $\mathbf{R}^d$-valued Brownian motion $(\Omega,\underline{\underline{F}},\{\underline{\underline{F}}_t\},P,\{w_t\})$, all of which may depend on u. Moreover this space is to carry a continuous adapted process $\{x_t\}$ satisfying

$$(3.2) \qquad \begin{aligned} &\text{(i)} && P\circ x_0^{-1} = \mu, \\ &\text{(ii)} && dx_t = f(t,x_t,u_t)dt + \sigma(t,x_t)dw_t, \\ &\text{(iii)} && E\int_0^T|x_t|^q dt < \infty, \quad E\int_0^T|u_t|^q dt < \infty, \end{aligned}$$

where $u_t = u(t,x)$.

For u in $\mathcal{U}_0$ we can define

$$(3.3) \qquad J_i(u) = E\{\int_0^T \ell_i(t,x_t,u_t)dt + c_i(x_T)\} \quad i = -m_1,\cdots,0,\cdots,m_2.$$

For any subset $\mathcal{U}\subset\mathcal{U}_0$ we can now define the problem as

$$(3.4) \qquad \inf\{J_0(u): u \text{ in } \mathcal{U},\ J_i(u) \le 0 \text{ if } i > 0,\ J_i(u) = 0 \text{ if } i < 0\}.$$

Let us say that u is admissible if u is in $\mathcal{U}$, $J_i(u) \le 0$ if $i > 0$, $J_i(u) = 0$ if $i < 0$.

We assume that u^* solves (3.4). Our aim is to establish conditions necessarily satisfied by u^*. From now on $(\Omega,\underline{\underline{F}},\{\underline{\underline{F}}_t\},P,\{w_t\},\{x_t\})$ are assumed to correspond to u^*.

3.2 Remark. The terminal time T could be replaced by $T\wedge\tau_A$ where τ_A is the first exit time from an open set A in $\mathbf{R}^n$ by simply multiplying ℓ_i,f,σ by $1_A(x)$, the characteristic function of A. Moreover the theory can also be extended to the case where ℓ_i, f, σ depend on the past of x.

To deduce necessary conditions we require comparison controls, so $\mathcal{U}$ must be sufficiently rich to yield "good" necessary conditions. We will perturb u^* to u, but we must ensure that u is in $\mathcal{U}$, so at least a corresponding solution of (3.2) must exist. Here we will use weak solutions defined via Girsanov's theorem. We begin with some hypotheses. Let

$$F(t,x) = \text{span}\{f(t,x,u) - f(t,x,v): u,v \text{ in } U\}.$$

If $\sigma(t,x)$: $\mathbf{R}^d \to \mathbf{R}^n$ is onto F(t,x), then for ξ in F(t,x)

$$\sigma(t,x)\theta = \xi$$

has a solution, and we shall always take it to be the one which has minimum norm, i.e. which lies in the range of σ(t,x)', the transpose of σ(t,x). Then ξ → θ is a linear operator which we denote by $\sigma(t,x)^+$. Since we shall also want to apply σ^+ to elements of $\mathbf{R}^n$ which may not lie in F(t,x), we shall in that case replace σ^+ by $\sigma^+\pi(t,x)$ where π(t,x) is the orthogonal projection of $\mathbf{R}^n$ onto F(t,x) and we shall do this without writing π(t,x) explicitly. The most common case is

$$f(t,x,u) = \begin{bmatrix} g(t,x) \\ h(t,x,u) \end{bmatrix}, \quad \sigma(t,x) = \begin{bmatrix} 0 \\ \sigma_2(t,x) \end{bmatrix}$$

with $\sigma_2(t,x)$ invertible, in which case

$$\pi(t,x)\xi = \begin{bmatrix} 0 \\ \xi_2 \end{bmatrix}, \quad \sigma(t,x)^+\xi = \sigma_2(t,x)^{-1}\,\xi_2.$$

We assume

(3.5) $$\int_0^T |\sigma(t,x)|^2 dt \le k_0, \quad |f(t,x,u)|^2 \le k_1(1+|x|^2+|u|^2)$$
$$\sigma(t,x): \mathbf{R}^d \to \mathbf{R}^n \text{ is onto } F(t,x), |\sigma(t,x)^+| \le k_0 \text{ for all } (t,x);$$

(3.6) there exists a constant K^* such that
$$|u^*(t,x)| \le K^*(1+\|x\|_t) \text{ a.s.};$$

(3.7) there exists ε > 0 such that $E\exp(\varepsilon|x_0|^2) < \infty$.

3.3 Lemma. Assume (3.5) - (3.7) and let $1 \le p < \infty$. Then $E\|x\|_T^p < \infty$.

Proof: Theorem 3.8.2 implies

$$E\left\|\int_0^\cdot \sigma(t,x_t)dw_t\right\|_T^p \le c_p^{-1}E\left(\int_0^T|\sigma|^2dt\right)^{p/2} \le k_0^{p/2}c_p^{-1}$$

so from lemma 2.1 it follows that for some K < ∞

$$E\|x\|_T^p \le K(1+E|x_0|^p) < \infty.$$

Q.E.D.

Let $\mathcal{U}_G$ be the set of measurable $\{\underline{\underline{G}}_t^n\}$-adapted functions $u:[0,T] \times C^n \to U$ such that for some constant K_u

(3.8) $$|u_t| \le K_u(1+\|x\|_t).$$

For u in $\mathcal{U}_G$ we can define x via Girsanov's theorem. Indeed for such u

(3.9) $$\theta_t^u = \sigma(t,x_t)^+[f(t,x_t,u_t) - f(t,x_t,u_t^*)],$$

and (3.5),(3.6),(3.8) imply that

(3.10) $$|\theta_t^u|^2 \le \bar{K}_u(1+\|x\|_t^2) \quad \text{a.s.}$$

for some $\bar{K}_u$. Corollaries 2.3, 2.4 and Girsanov's theorem imply that

$$dx = f(t,x_t,u_t)dt + \sigma(t,x_t)dw_t^u$$

and

(3.11) $$E\{\zeta_T(\theta^u)^p\} \le k_u$$

for some p > 1 depending on $\bar{K}_u$, where

$$w_t^u = w_t - \int_0^t \theta_s^u \, ds$$

is a Brownian motion on $(\Omega,\underline{\underline{F}},\{\underline{\underline{F}}_t\},P^u)$ and $dP^u = \zeta_T(\theta^u)dP$. Moreover if $p' = p/(p-1)$ then

$$E^u\int_0^T|x_t|^q dt = E\{\zeta_T(\theta^u)\int_0^T |x_t|^q dt\}$$

$$\le k_u^{1/p} \; T \; E\{\|x\|_T^{qp'}\}^{1/p'}$$

is finite by lemma 3.3. Now (3.8) implies

$$E^u \int_0^T|u_t|^q dt \le T\, k_u^{1/p}\, K_u^q\, E\{(1+\|x\|_T)^{p'q}\}^{1/p'} < \infty$$

so that u is in $\mathcal{U}_0$ and we have established

3.4 <u>Lemma</u>. Assume (3.5) - (3.7). Then $\mathcal{U}_G \subset \mathcal{U}_0$ and u^* is in $\mathcal{U}_G$.

From lemma 3.4 we see that we can take for comparison controls any element of $\mathcal{U}\cap\mathcal{U}_G$. As the corresponding "state" we cannot really take $\{x_t\}$, since this does not change in the Girsanov setting, but rather we ought to take $\zeta_t(\theta^u)$. According to (1.3), $\zeta_t(\theta^u)$ satisfies

$$d\zeta_t = \zeta_t \theta_t^u \cdot dw_t$$

which equation would constitute the state equation. But now we face a difficulty. We shall see shortly that the method of strong perturbations consists of taking a comparison control u_t^ε which is equal to u_t^* except for t in I_ε, an interval of length ε, and it is required that such a perturbation gives rise to a perturbation of the cost $J(u^*)$ of order ε. Roughly speaking for this one parameter family of perturbed controls $\{u^\varepsilon\}$, the necessary condition is $\frac{d}{d\varepsilon} J(u^\varepsilon)\big|_{\varepsilon=0} \geq 0$. But a prerequisite for the indicated derivative to exist is that $J(u^\varepsilon) - J(u^*) = O(\varepsilon)$. If we denote perturbations by δ, then

$$\delta u = O(\varepsilon) \Rightarrow \delta\theta = O(\varepsilon) \Rightarrow \delta\zeta = O(\varepsilon^{1/2})$$

since $(dw)^2 = dt$. Now we only obtain

$$\delta J = O(\varepsilon^{1/2}).$$

Hence for the method to work, we must not have stochastic integrals involving u. We now massage the problem to the point where this is the case.

For u in $\mathcal{U}_G$ we have

$$\begin{aligned} J(u) &= E^u\{\int_0^T \ell(t,x_t,u_t)dt + c(x_T)\} \\ &= E^u\{\int_0^T [\ell(t,x_t,u_t) - \ell(t,x_t,u_t^*)]dt \\ &\qquad + \int_0^T \ell(t,x_t,u_t^*)dt + c(x_T)\}. \end{aligned}$$

Let

$$\phi_t^u = \ell(t,x_t,u_t) - \ell(t,x_t,u_t^*)$$

$$L = \int_0^T \ell(t,x_t,u_t^*)dt + c(x_T)$$

so that

(3.12) $$J(u) = E^u\int_0^T \phi_t^u dt + E^u L.$$

Now (3.1), (3.5) - (3.7) imply that $E|L|^2 < \infty$ so that $L_t \equiv E\{L|\bar{\underline{F}}_{t+}\}$ is a square integrable martingale. A martingale decomposition theorem, cf. appendix 3.8.3, implies that there exists a measurable adapted process $\{\chi_t\}$ such that

(3.13) $$\int_0^T E|\chi_t|^2 dt < \infty,$$

as well as a square integrable martingale $\{M_t\}$, "orthogonal" to $\{w_t\}$ with $M_0 = 0$, such that

(3.14) $$L_t = L_0 + \int_0^t \chi_s dw_s + M_t \quad \text{a.s.}$$

We can now show that $\{M_t\}$ is irrelevant.

3.5 Lemma. $E^u L = EL + E^u \int_0^T \chi_t \theta_t^u dt$ if u is in $\mathcal{U}_G$.

Proof: From (3.14) and (1.6) we obtain

$$E^u L = EL + E^u \int_0^T \chi dw^u + E^u \int_0^T \chi\theta^u dt + E^u M_T.$$

According to lemma 3.8.4

$$E^u \int_0^T |\chi|^2 dt < \infty$$

so that

$$E^u \int_0^T \chi dw^u = 0,$$

and according to lemma 3.8.5, $E^u M_T = 0$; hence the result follows.

Q.E.D.

The two lemmas and (3.12) now imply that u^* is also a solution of

(3.15) $\inf\{\tilde{J}_0(u) : u \text{ in } \mathcal{U}\cap\mathcal{U}_G,\ \tilde{J}_i(u) \leq 0 \text{ if } i > 0,\ \tilde{J}_i(u) = 0 \text{ if } i < 0\}$

where

(3.16) $$\tilde{J}(u) = E^u \int_0^T (\phi_t^u + \chi_t \theta_t^u) dt + J(u^*),$$

(3.17) $$\phi_t^u = \ell(t,x_t,u_t) - \ell(t,x_t,u_t^*)$$

(3.18) $$\theta_t^u = \sigma(t,x_t)^+ [f(t,x_t,u_t) - f(t,x_t,u_t^*)].$$

It is to this problem that we shall apply a Lagrange multiplier rule.

3.6 Remark. It is clear that the more general constraint $E\tilde{L}(x) \leq 0$ $(= 0)$ could be included. It is treated just as L is treated in (3.12), provided $\tilde{L}$ is a Borel measurable function defined on the continuous R^n-valued functions such that $|\tilde{L}(x)| \leq k_2(1+\|x\|_T^q)$.

3.7 Remark. If we wish to consider the Markov controls we take $\mathcal{U} = \{u \text{ in } \mathcal{U}_0 : u_t = \psi(t,x_t),\ \psi:[0,T]\times\mathbf{R}^n \to U \text{ is Borel measurable}\}$. Now (3.6) becomes $|u^*(t,x)| \leq K^*(1+|x|)$. Moreover $\mathcal{U}\cap\mathcal{U}_G$ corresponds to U-valued Borel measurable functions ψ such that $|\psi(t,x)| \leq K_\psi(1+|x|)$.

3.8 Appendix. We shall state here some technical results from the general theory of stochastic processes, and then we shall establish two lemmas required in the proof of lemma 3.5.

Given $(\Omega,\underline{\underline{F}},\{\underline{\underline{F}}_t\}_{0\leq t\leq T},P)$ we say that the predictable σ-algebra is the σ-algebra $\underline{\underline{P}}$ on $[0,T] \times \Omega$ generated by the adapted processes $\{x_t\}$ with left-continuous trajectories. We observe that if we change $\underline{\underline{F}}_t$ into $\underline{\underline{F}}_{t+}$ then $\underline{\underline{P}}$ does not change. Then we say that the process $\{x_t\}$ is predictable if $(t,\omega) \to x_t(\omega)$ is $\underline{\underline{P}}$-measurable. It follows that a predictable process is adapted.

If $\{x_t\}$ is an $\{\bar{\underline{\underline{F}}}_{t+}\}$ martingale, then it can be modified to have right continuous trajectories with left limit, and the jump at any time t,

$$\Delta x_t = x_t - x_{t-},$$

is well defined. Let $\{x_t\}$, $\{y_t\}$ be two $\mathbf{R}^n$-valued square integrable $\{\bar{\underline{\underline{F}}}_{t+}\}$ martingales. Then $[x,y]_t$ is the unique $\mathbf{R}^n\otimes\mathbf{R}^n$-valued process of bounded variation such that $x_t y_t' - [x,y]_t$ is an $\{\bar{\underline{\underline{F}}}_{t+}\}$ martingale and $\Delta[x,y]_t = \Delta x_t \Delta y_t'$. Moreover $\langle x,y\rangle_t$ is the unique $\{\bar{\underline{\underline{F}}}_{t+}\}$ predictable process such that $x_t y_t' - \langle x,y\rangle_t$ is an $\{\bar{\underline{\underline{F}}}_{t+}\}$ martingale, with $x_0 y_0 - \langle x,y\rangle_0 = 0$. We write $\langle x\rangle_t$ for $\langle x,x\rangle_t$ and $[x]_t$ for $[x,x]_t$. If $\{x_t\}$ is left-continuous then it is predictable and $[x]_t = \langle x\rangle_t$. (Since the filtration is right continuous then $\{x_t\}$ is in fact continuous.) Otherwise

$$\langle x\rangle_t - \sum_{s\leq t} E\{\Delta x_s(\Delta x_s)'|\bar{\underline{\underline{F}}}_{s-}\}$$

$$= [x]_t - \sum_{s\leq t} \Delta x_s(\Delta x_s)'.$$

Note that $\bar{\underline{\underline{F}}}_{s-}$ is the smallest σ-algebra containing $\bar{\underline{\underline{F}}}_r$ for all $r < s$. Finally $\{x_t\}$, $\{y_t\}$ are orthogonal if $\langle x,y\rangle_t = 0$ for all t.

3.8.1 Remark. If

$$x_t = \int_0^t \sigma_s \, dw_s$$

then

$$[x]_t = \langle x\rangle_t = \int_0^t \sigma_s \, \sigma_s' ds.$$

3.8.2 Theorem (Burkholder-Davis-Gundy). For any p such that $1 \leq p < \infty$, there exist constants c_p, C_p depending on p and n, such that for any $\mathbf{R}^n$ -valued martingale $\{x_t\}$,

$$c_p \; E\|x\|_T^p \leq E\{|[x]_T|^{p/2}\} \leq C_p \; E\|x\|_T^p.$$

The proof for the case n=1 can be found in Dellacherie and Meyer (1980), chapter VII, §92. The extension to n>1 is trivial if we recall

$$[x,y] = 1/4\{[x+y] - [x-y]\}.$$

For the case of stochastic integrals we have

$$c_p \; E \left\|\int_0^\bullet \sigma dw\right\|_T^p \leq E\{(\int_0^T|\sigma|^2 dt)^{p/2}\} \leq C_p \; E \left\|\int_0^\bullet \sigma dw\right\|_T^p$$

since $|[\int_0^\bullet \sigma dw]_T|$ and $\int_0^T|\sigma|^2 dt$ are equivalent.

3.8.3 Theorem (Kunita-Watanabe). Let $\{w_t\}$ be a Brownian motion on $(\Omega, \underline{\underline{F}}, \{\underline{\underline{F}}_t\}, P)$. If $\{x_t\}$ is a square integrable $\mathbf{R}^n$-valued $\{\bar{\underline{\underline{F}}}_{t+}\}$ martingale, then there exists a $\{\bar{\underline{\underline{F}}}_{t+}\}$ predictable process $\{\chi_t\}$ and a square integrable martingale $\{m_t\}$, orthogonal to $\{w_t\}$ such that

$$\int_0^T E|\chi_t|^2 dt < \infty,$$

$$x_t = \int_0^t \chi_s dw_s + m_t.$$

If $\{x_t\}$ is $\{\underline{\underline{F}}_t^w\}$ adapted then $\{m_t\} = 0$.
The proof can be found in Dellacherie and Meyer (1980), chapter VIII, §48, 49. We remark that we used the fact that $\{w_t\}$ is also an $\{\bar{\underline{\underline{F}}}_{t+}\}$ Brownian motion.

We observe that since $\{\chi_t\}$ is $\{\bar{\underline{\underline{F}}}_{t+}\}$ predictable, then a.s it is $\{\underline{\underline{F}}_t\}$ predictable, cf. Dellacherie and Meyer (1980), appendix 1, §7, hence we can replace $\{\chi_t\}$ by this $\{\underline{\underline{F}}_t\}$ adapted version in (3.14).

In the proof of the following lemma we appeal to a result of Doob which we recall here: if $\{m_t\}$ is a martingale, $p > 1$, then

$$E\|m\|_T^p \leq c \; E|m_T|^p$$

for some constant c. For a proof we refer the reader to Doob (1953), pages 317, 354.

3.8.4 Lemma. $E^u \int_0^T |\chi|^2 dt < \infty$

Proof: The point here is that although we have (3.13), we need this with E replaced by E^u. Let $p' > 1$. Then from (3.14) and theorem 3.8.2 it follows that

$$E\{[\int_0^T |\chi|^2 dt + |[M]_T|]^{p'}\} = E\{|[L. - L_0]_T|^{p'}\}$$

(3.19)
$$\leq C_1 \; E\{\|L. - L_0\|_T^{2p'}\}$$

$$\leq C_2 \; E|L_T - L_0|^{2p'}$$

by Doob's inequality. The last expression is finite because

$$|L_T| = |L| \leq k_2\{\int_0^T(1+|x_t|^q + |u_t|^q)dt + |x_T|^q\}$$

so (3.6) and lemma 3.3 give a finite bound. But now let $p' = p/(p-1)$ with p as in (3.11), so that

$$E^u\int_0^T|\chi|^2 dt = E\{\zeta_T(\theta^u)\int_0^T|\chi|^2 dt\}$$

$$\leq E\{\zeta_T(\theta^u)^p\}^{1/p} \; E\{(\int_0^T|\chi|^2 dt)^{p'}\}^{1/p'}$$

$$\leq k_u^{1/p}\{C_2 E|L_T - L_0|^{2p'}\}^{1/p'} < \infty$$

by (3.11) and (3.19). Q.E.D.

3.8.5 Lemma. $E^u M_T = 0$.

Proof: Again we know that $EM_T = EM_0 = 0$, but we require the result under E^u. Let

$$\tau_N = \inf\{t \geq 0: |\int_0^t \theta^u \cdot dw| \geq N\}.$$

Then

$$E\{\zeta_{T\wedge\tau_N}(\theta^u)M_T\} = E\{M_T + \int_0^T 1_{\tau_N} \zeta_t(\theta^u)\theta_t^u \cdot dw \; M_T\} = 0.$$

Indeed

$$\int_0^T E\{1_{\tau_N} \; \zeta_t(\theta^u)^2|\theta^u|^2\}dt$$

$$\leq e^{2N} \; \bar{K}_u T(1+E\|x\|_T^2) < \infty$$

so that

$$I(t) = \int_0^t 1_{\tau_N} \; \zeta_s(\theta^u) \; \theta_s^u \cdot dw_s$$

is a square integrable martingale. Moreover we can replace θ^u by its predictable projection without changing $I(t)$, a.s. Thus $I(t)$, as the stochastic integral of a predictable process, is orthogonal to M_t, i.e. $E\{I(T)M_T\} = E\{I(0)M(0)\} = 0$.

The continuity of $t \to \zeta_t(\theta^u)$ implies that

$$\zeta_{T\wedge\tau_N}(\theta^u)M_T \to \zeta_T(\theta^u)M_T \qquad \text{a.s.}$$

and for $1 < \bar{p} < p$

$$E|\zeta_{T\wedge\tau_N}(\theta^u)M_T|^{\bar{p}} \leq E\{\zeta_T(1_{\tau_N}\theta^u)^p\}^{\bar{p}/p}E\{|M_T|^{p\bar{p}/(p-\bar{p})}\}^{\frac{p-\bar{p}}{p}}$$

$$\leq k_u^{\bar{p}/p}c_3E\{|[M]_T|^{p\bar{p}/2(p-\bar{p})}\}^{\frac{p-\bar{p}}{p}}$$

$$< \infty$$

by (3.19) with $p' = p\bar{p}/2(p-\bar{p})$. Note that we have used (3.11) and that p, k_u which depend on $(1_{\tau_N}\theta^u)$ can be taken as those generated by θ^u, i.e. independent of N since $\bar{K}_u$ is uniform in N. It now follows that

$$0 = E\{\zeta_{T\wedge\tau_N}(\theta^u)M_T\} \to E\{\zeta_T(\theta^u)M_T\} = E^uM_T.$$

Q.E.D.

3.9 Comments. The material of this section appeared originally in Haussmann (1976), where in addition the data are allowed to depend on the past of x.

4 The abstract multiplier theorem

We wish to establish a Langrange multiplier rule for the problem

(4.1) $\inf\{\tilde{J}_0(u): u \text{ in } \mathcal{U}, \tilde{J}_i(u) \leq 0, \tilde{J}_j(u) = 0, i > 0, j < 0\}$

where $\tilde{J}_\ell : \mathcal{U} \to \mathbf{R}$, $\ell = -m_1, \cdots, 0, \cdots, m_2$ and $\mathcal{U}$ is an arbitrary set. We require some preliminaries.

4.1 Definition. The set X (a topological space) has the fixed point property if for each continuous function $f: X \to X$ there exists x in X such that $f(x) = x$.

4.2 Brouwer Fixed Point Theorem. The closed unit sphere of $\mathbf{R}^n$ has the fixed point property.

The proof can be found in Dunford and Schwartz (1965), V.12. It follows as a corollary that any closed convex set in $\mathbf{R}^n$ has the fixed point property. If n = 1 the theorem is self-evident from diagram 4.1. Indeed the closed convex set is the interval [a,b], and the graph of a continuous function must evidently cross the diagonal at least once, hence yield a fixed point.

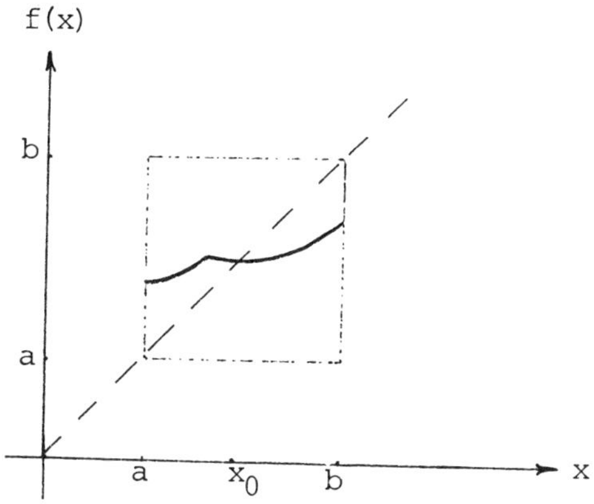

Diagram 4.1

4.3 Eidelheit Separation Theorem. Let A, B be convex sets in a topological vector space. If int A $\neq \emptyset$ and B $\cap$ int A = $\emptyset$, then there exists a separating hyperplane, i.e. a non-zero continuous linear functional v, and a constant c, such that

$$\sup_{x \in B} v(x) \leq c \leq \inf_{y \in A} v(y).$$

Here int A denotes the interior of the set A. The proof can again be found in Dunford and Schwartz (1965), V.2.

Now we need a generalization of directional derivatives. Let C be a standard simplex in $\mathbf{R}^{\ell}$, i.e.

$$C = \{(\rho_1,\cdots,\rho_\ell): \rho_i \geq 0, \sum_1^{\ell} \rho_i \leq 1\}$$

4.4 Definition. Given u in $\mathcal{U}$, $\eta > 0$, $z:\eta C \to \mathcal{U}$ with $z(0) = u$, then $M: \mathbf{R}^{\ell} \to \mathbf{R}^{m_1+1+m_2}$ is the conical differential of $\tilde{J}$ along z at u if

$$\tilde{J}(z(\rho)) = \tilde{J}(u) + M\rho + o(|\rho|). \tag{4.2}$$

Here $o(|\rho|)$ is an element of $\mathbf{R}^{m_1+1+m_2}$ such that $|o(|\rho|)|/|\rho| \to 0$ as $|\rho| \to 0$. Observe that if $\ell = 1$, $z(\rho) = u + \rho v$ and if $\tilde{J}$ is Gateaux differentiable with derivative denoted by $\delta\tilde{J}$, then $M = \delta\tilde{J}(u;v)$. This example is expanded upon in exercises 4.8.1, 4.8.2, 4.8.3.

4.5 Definition. A convex cone $D \subset \mathbf{R}^{m_1+1+m_2}$ with vertex at 0 is a cone of variations of $\tilde{J}$ at u if for all linearly independent $\{d^1,d^2,\cdots,d^{\ell}\} \subset D$ there exists $\eta > 0$ and $z:\eta C \to \mathcal{U}$, $z(0) = u$, such that $\tilde{J}\circ z$ is continuous and $M = (d^1,\cdots,d^{\ell})$ is the conical differential of $\tilde{J}$ along z at u.

Fix u^* in $\mathcal{U}$. Let us write $I_- = \{i > 0: \tilde{J}_i(u^*) < 0\}$, $I_+ = \{i > 0: \tilde{J}_i(u^*) = 0\}$, $I_i = +\infty$ if i in I_-, $I_i = 0$ if i in I_+.

Let us also set $L = \{y \text{ in } \mathbf{R}^{1+m_2}: y_0 < 0, y_i < I_i, i=1,\cdots,m_2\}$.

4.6 Lemma. If u^* solves (4.1) and if D is a cone of variations of $\tilde{J}$ at u^*, then there exists $\tilde{\lambda}$ in $\mathbf{R}^{m_1+1+m_2}$, $\tilde{\lambda} \neq 0$, such that $\tilde{\lambda}\cdot d \leq 0$ for all d in $D \cap (\mathbf{R}^{m_1} \times L)$.

Proof: Let $\underline{D} = \{y \text{ in } \mathbf{R}^{m_1} : y_i = d_{-i},\ i=1,\cdots,m_1, d \text{ in } D \cap (\mathbf{R}^{m_1}\times L)\}$. If we take $\tilde{\lambda} = (\tilde{\lambda}_-, 0,\cdots,0)$, $\tilde{\lambda}_-$ in $\mathbf{R}^{m_1}$, then the lemma simply claims that 0 and $\underline{D}$ are separated. We first establish that 0 is not in int $\underline{D}$. Suppose the opposite, i.e. 0 is in int $\underline{D}$. Then there is a ball centred at 0 contained in $\underline{D}$. Let S be an m_1-simplex contained in the ball such that 0 is in int S, and let $y^0,\cdots,y^{m_1}$ be the vertices of S. Then $0 = \sum_{i=1}^{m_1} \bar{\rho}_i\, y^i$ for a unique set $\{\bar{\rho}_i\}$, and moreover $\bar{\rho}_i > 0$. Let $\bar{d}^0,\cdots,\bar{d}^{m_1}$ in $D \cap (\mathbf{R}^{m_1}\times L)$ correspond to these vertices, i.e. if we define the projections π and π_0 by

$$\pi(y_{-m_1},\cdots,y_0,\cdots,y_{m_2}) = (y_{-m_1},\cdots,y_0)$$

$$\pi_0(y_{-m_1},\cdots,y_{-1},y_0) = (y_{-m_1},\cdots,y_{-1})$$

then $\pi_0\pi\bar{d}^i = y^i$. Since $\bar{d}^i_0 < 0$ for all i (cf. the definition of L) and since $\bar{\rho}_i > 0$ (cf. above) then $\{\bar{d}^i\}_0^{m_1}$ are independent. Since $D\cap(\mathbf{R}^{m_1}\times L)$ is a convex cone we can scale $\bar{d}^i$ suitably to obtain $\{d^i\}_0^{m_1} \subset D$, linearly independent, such that for all i $d^i_0 = -1$, $d^i_j < I_j$ if $j > 0$, (i.e. $d^i = \bar{d}^i|\bar{d}^i_0|^{-1}$), and $(0,0,\cdots,0,-1)$ is in the relative interior of $\pi\mathcal{S}$ where $\mathcal{S}$ is the simplex with vertices $\{d^0,\cdots,d^{m_1}\}$. Moreover, by construction $\{\pi d^i\}_0^{m_1}$ are linearly independent. The reader should consult diagram 4.2. Let $M = (d^0,d^1,\cdots,d^{m_1})$. Since D is a cone of variations, there exist $\eta > 0$, $z: \eta C \to \mathcal{U}$, such that (4.2) holds. Let $\Delta = \text{co}\{0,d^0,\cdots,d^{m_1}\}$, the convex hull of $\{0,d^0,\cdots,d^{m_1}\}$, $\Delta_\pi = \pi\Delta$, $M_\pi = \pi\circ M$. Then $M: \eta C \to \eta\Delta$ is onto as is $M_\pi: \eta C \to \eta\Delta_\pi$ so that M_π^{-1} exists. Define $\psi: \eta\Delta_\pi \to \mathbf{R}^{m_1+1}$ by

$$\psi(y) = \pi\tilde{J}\circ z\circ M_\pi^{-1}(y) - \pi\tilde{J}(u^*).$$

Note

A typical example with $m_1 = m_2 = 1$

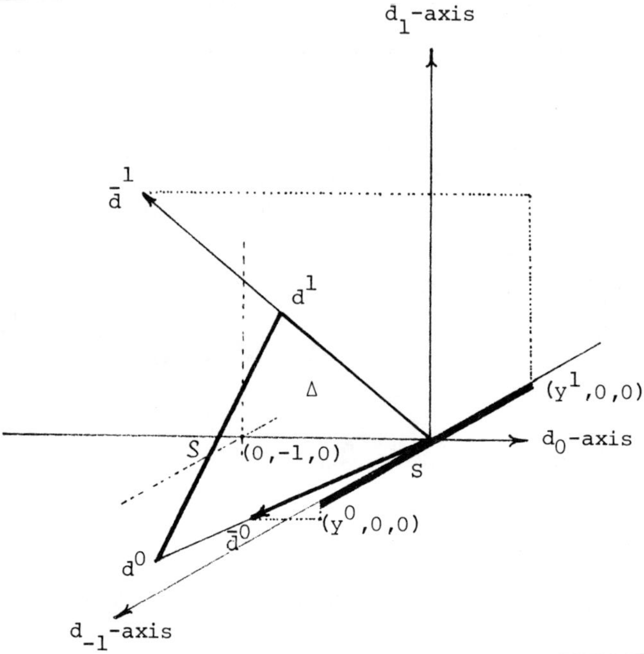

d^0, d^1 lie in the plane $\{d_0 = -1\}$.

S is the line segment $[(y^0,0,0)\ (y^1,0,0)]$.

S is the line segment $[d^0,d^1]$.

Δ is the triangle $[0,d^0,d^1]$.

The image under π

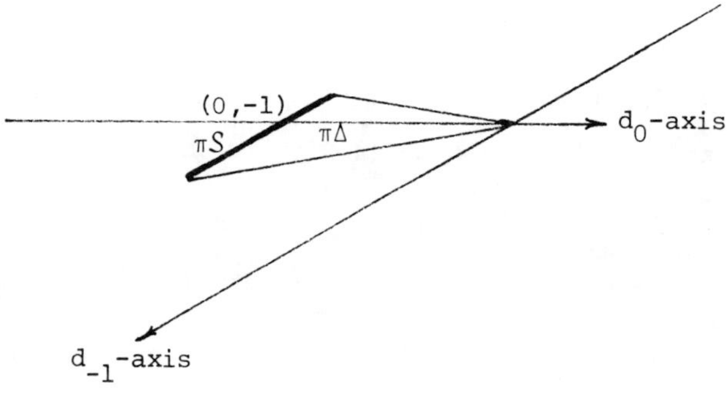

π projects out the $\mathbb{R}^{m_2}$ components, i.e. the d^1 component here.

Diagram 4.2

$$(4.3)\qquad \psi(y) = M_\pi(M_\pi^{-1}y) + o(|M_\pi^{-1}y|) = y + o(|M_\pi^{-1}y|).$$

Let $a = (0,0,\cdots,-1/2)$ in int Δ_π, $\varepsilon_0 = \min\{1,|d_j^i|;i=0,\cdots,m_1,j \text{ in } I_+\}$. Then there exists ε in $(0,\varepsilon_0)$ such that $S_\varepsilon(a)$, the ball of radius ε centred at a, is contained in int Δ_π. Moreover from the definition of o in (4.3) there exists $\delta_0 > 0$ such that $|o(|M_\pi^{-1}y|)| < \varepsilon|M_\pi^{-1}y|$ if $|M_\pi^{-1}y| < \delta_0$. Let $\delta = \min\{\delta_0, \eta, |\tilde{J}_j(u^*)|[\varepsilon + \max_i|d_j^i|]^{-1}: j \text{ in } I_-\}$. Now if y is in $\delta\Delta_\pi$ then $M_\pi^{-1}y$ is in δC (because $M_\pi^{-1}\Delta_\pi = C$) so $|M_\pi^{-1}y| \leq \delta$ and hence from (4.3), if y is in $\delta\Delta_\pi$ then

$$(4.4)\qquad |o(|M_\pi^{-1}y|)| \leq \varepsilon\delta, \text{ i.e.} |\psi(y) - y| \leq \varepsilon\delta.$$

Next $\pi_0 a = 0$ so $\pi_0\, S_\varepsilon(a) = S'_\varepsilon(0) \subset \pi_0\Delta_\pi$ and

$$(4.5)\qquad \overline{S'_{\varepsilon\delta}(0)} = \delta\overline{S'_\varepsilon(0)} \subset \delta\pi_0\Delta_\pi = \pi_0\ \delta\Delta_\pi,$$

where the overbar denotes closure. Define Φ: $\pi_0(\delta\Delta_\pi) \to \mathbf{R}^{m_1}$ by

$$\Phi(y) = y - \pi_0\psi\big((y,-\delta)\big) = \pi_0[(y,-\delta) - \psi\big((y,-\delta)\big)].$$

We now show that Φ has a fixed point. Since $\pi_0(\delta\Delta_\pi)$ is a simplex (i.e. closed, convex) then the Brouwer fixed point theorem is applicable. Φ is continuous since π_0, π, $\tilde{J}\circ z$, and M_π^{-1} are. If y is in $\pi_0(\delta\Delta_\pi)$ then $(y,-\delta)$ is in $\delta\Delta_\pi$ and by (4.4)

$$(4.6)\qquad |\psi\big((y,-\delta)\big) - (y,-\delta)| \leq \varepsilon\delta$$

so $|\Phi(y)| \leq |\pi_0|\varepsilon\delta = \varepsilon\delta$. From (4.5) $\Phi(\pi_0\delta\Delta_\pi)$ $\overline{S'_{\varepsilon\delta}(0)} \subset \pi_0(\delta\Delta_\pi)$. Hence by theorem 4.2, Φ has a fixed point, call it $\bar{y}$, in $\pi_0\delta\Delta_\pi$ and $(\bar{y},-\delta)$ is in $\delta\Delta_\pi$. Let y_0 be given by

$$(4.7)\qquad \psi\big((\bar{y},-\delta)\big) = (0,\cdots,0,y_0).$$

From (4.6),(4.7)

$$\varepsilon\delta \geq |\psi(\bar{y},-\delta) - (\bar{y},-\delta)| = |(\bar{y},y_0+\delta)| \geq |y_0+\delta|$$

so that $y_0 \leq \varepsilon\delta - \delta < 0$ since $\varepsilon < 1$. Hence from (4.7) and the definition of ψ it follows that

$$0 > y_0 = \tilde{J}_0[z_\circ M_\pi^{-1}(y,-\delta)] - \tilde{J}_0(u^*),$$

$$0 = \tilde{J}_i[z \circ M_\pi^{-1}(\bar{y},-\delta)] \text{ if } i < 0,$$

i.e. $z \circ M_\pi^{-1}(\bar{y},-\delta)$ is in $\mathcal{U}$, gives a lower cost and satisfies the equality constraints. Call this element u in $\mathcal{U}$. Since $(\bar{y},-\delta)$ is in $\delta\Delta_\pi$ and since $d_0^i = -1$, then for some ρ

$$(\bar{y},-\delta) = \pi \sum_{i=0}^{m_1} \rho_i d^i, \; \rho_i \geq 0, \; \sum \rho_i = \delta,$$

so from (4.2)

$$\tilde{J}_j(u) - \tilde{J}_j(u^*) = \sum_{i=0}^{m_1} \rho_i d_j^i + o(|M_\pi^{-1}(\bar{y},-\delta)|).$$

If j is in I_- then from (4.4) and the definition of δ it follows that

$$|\tilde{J}_j(u) - \tilde{J}_j(u^*)| \leq \delta \max_i |d_j^i| + \varepsilon\delta \leq -\tilde{J}_j(u^*)$$

so that $\tilde{J}_j(u) \leq 0$, and if j is in I_+ then from the definition of ε, (4.4), and the fact that $d_j^i < 0$, we have

$$\tilde{J}_j(u) \leq \delta \max_i d_j^i + \varepsilon\delta < 0.$$

Note that the definitions of $\underline{D}$ and I_+ imply that $d_j^i < 0$ for all i if j is in I_+. Hence u also satisfies the inequality constraints and we have a contradiction. Thus 0 is not in int $\underline{D}$. If int $\underline{D} = \emptyset$ then $\underline{D}$ lies in an m_1-1 dimensional hyperplane which then separates 0 and $\underline{D}$ cf. exercise 4.8.4. Finally if int $\underline{D} \neq \emptyset$ then theorem 4.3 gives the result.

Q.E.D.

4.7 Theorem. If u^* solves (4.1) and if D is a cone of variations of $\tilde{J}$ at u^* then there exists λ in $\mathbf{R}^{m_1+1+m_2}$, $\lambda \neq 0$, $\lambda_i \leq 0$ if $i \geq 0$ such that $\lambda \cdot \tilde{J}(u^*) = \lambda_0 \tilde{J}_0(u^*)$ and $\lambda \cdot d \leq 0$ for d in D.

Proof: With $\tilde{\lambda}$ as in lemma 4.6 set

$$B = \{y \text{ in } \mathbf{R}^{2+m_2} : y_{-1} = \tilde{\lambda}\cdot d, \; y_i \geq d_i, \; i \geq 0, \; d \text{ in } D\},$$

$$A = \{y \text{ in } \mathbf{R}^{2+m_2} : y_{-1} > 0, \; (y_0,\cdots,y_{m_2}) \text{ in } \;\; \}.$$

We write y_+ for $(y_0,\cdots,y_{m_2})$. Then A, B are convex, int A = A $\neq \emptyset$, and

$A \cap B = \emptyset$ by lemma 4.6. Hence there exists λ_+ in $\mathbf{R}^{1+m_2}$, and η_1, η_2 such that $(\eta_2, \lambda_+) \neq 0$

$$\eta_2 \tilde{\lambda} \cdot d + \lambda_+ \cdot y_+ \leq \eta_1 \tag{4.8}$$

for $y_+ \geq d_+$, d in D, and

$$\eta_1 \leq \eta_2 \xi + \lambda_+ \cdot \bar{y}_+ \tag{4.9}$$

for $\xi > 0$, $\bar{y}_+$ in L. By continuity this last inequality holds for $\xi \geq 0$ and $\bar{y}_+$ in $\bar{L}$. Recall that $\tilde{\lambda} = (\tilde{\lambda}_-, 0, \cdots, 0)$ and set $\lambda = (\eta_2 \tilde{\lambda}_-, \lambda_+)$. Then it follows that $\lambda \neq 0$ and for any d in D, y_+ in $\bar{L}$,

$$\lambda \cdot d \leq \eta_1 \leq \lambda_+ \cdot y_+ \tag{4.10}$$

by taking $y_+ = d_+$ in (4.8) and $\xi = 0$ in (4.9). Since 0 is in $\bar{L}$ then $\eta_1 \leq 0$ so $\lambda \cdot d \leq 0$. Moreover $\bar{L}$ contains the negative orthant so unless $\lambda_+ \leq 0$ there is no lower bound on $\lambda_+ \cdot y_+$ contradicting (4.10). Hence $\lambda_i \leq 0$ for $i \geq 0$. Finally if $y_+^* = -\left(0, \tilde{J}_1(u^*), \cdots, \tilde{J}_{m_2}(u^*)\right)$ then for any $\nu \geq 0$, νy_+^* is in $\bar{L}$, so

$$\eta_1 \leq \nu \lambda_+ \cdot y_+^* = -\nu \sum_{i>0} \lambda_i \tilde{J}_i(u^*).$$

Since ν can be arbitrarily large and since $\lambda_i \tilde{J}_i(u^*) \geq 0$ for $i > 0$ then $\sum_{i>0} \lambda_i \tilde{J}_i(u^*) = 0$ which establishes $\lambda \cdot \tilde{J}(u^*) = \lambda_0 \tilde{J}_0(u^*)$. Q.E.D.

As the results of this section are somewhat complicated the reader is encouraged to study the diagrams and to do the exercises.

4.8 Exercises

4.8.1 Assume that $\mathcal{U}$ is a convex set in a topological vector space. Assume also that $\tilde{J}$ is continuously Gateaux differentiable at u^*, i.e. for any v, any u near u^*

$$\tilde{J}(u+tv) - \tilde{J}(u) = \delta\tilde{J}(u;v)t + R(t,u,v), \quad t \geq 0,$$

where $|R(t,u,v)| = o(|t|)$, and for any v, $\delta\tilde{J}(\cdot;v)$ is continuous at u^*. Let $\{v^1,\cdots,v^\ell\}$ be such that $u^* + v^i$ is in $\mathcal{U}$ for any $i = 1,2,\cdots,\ell$, and set

$$z(\rho) = u^* + \sum_{i=1}^{\ell} \rho_i v^i$$

Show that the conical differential of $\tilde{J}$ along z at u^* is the matrix with columns $\delta\tilde{J}(u^*;v^i)$.

4.8.2 Assume that $\mathcal{U}$ is as above and that $\tilde{J}$ is continuously Gateaux differentiable at u^*. Let D be the convex cone generated by

$$\{\delta\tilde{J}(u^*;v-u^*): v \text{ in } \mathcal{U}\}$$

Show that D is a cone of variations of $\tilde{J}$ at u^*.

4.8.3 Let $\mathcal{U}$, $\tilde{J}$, D be as in exercise 4.8.2 and let u^* solve the problem (4.1). Show that there exists a non-zero λ in $\mathbf{R}^{m_1+1+m_2}$ such that

i) $\lambda_0 \le 0$, and for $i \ne 0$, $\lambda_i \tilde{J}_i(u^*) = 0$,

ii) $\lambda \cdot \delta\tilde{J}(u^*;v-u^*) \le 0$ for all v in $\mathcal{U}$.

4.8.4 Let K be a convex cone in $\mathbf{R}^d$ with vertex at 0 and let V be the subspace spanned by K. Show that if int K = ∅ then the dimension of V is less than d.

4.8.5 (The Lagrange multiplier rule of calculus) Let f and g be continuously differentiable functions defined on $\mathbf{R}^n$, $n \ge 2$. Use theorem 4.7 to show that there exists λ in $\mathbf{R}^2$, $\lambda \ne 0$, such that if x solves

$$\inf\{f(x): x \text{ in } \mathbf{R}^n,\ g(x) = 0\}$$

then $\lambda_0 \nabla f(x) + \lambda_{-1} \nabla g(x) = 0$.

4.8.6 Observe that $(1,\sqrt{3}/2)$ solves

$$\inf\{x^2+y^2: 4-x-2\sqrt{3}y \le 0,\ (x,y) \text{ in } \mathcal{U}\}$$

where $\mathcal{U} = \{(x,y): x^2 + 4y^2 - 4 \le 0\}$. Show that in this case $\lambda_0 = 0$. This implies that the constraint already determines $u^* = (1,\sqrt{3}/2)$ and so $\tilde{J}_0$ is irrelevant. Hint: the line $x + 2\sqrt{3}y = 4$ is tangent to $\mathcal{U}$.

4.8.7 Let $\tilde{J}(x,y) = \tilde{J}_0(x,y) = \begin{cases} x^{-1}(x^2+y^{3/2}) & \text{if } x,y > 0 \\ 0 & \text{if } x = y = 0 \end{cases}$

$\mathcal{U} = \{(x,y): x^2 \leq y \leq 2x^2, \quad x \geq 0\}$

It is clear that

$$\inf\{\tilde{J}(u): u \text{ in } \mathcal{U}\}$$

is attained at $u^* = (0,0)$.

i) Show that (1,1) is the conical differential of $\tilde{J}$ along z at (0,0) if

$$z(\rho_1,\rho_2) = (\rho_1+\rho_2, \; \rho_1^2+2\rho_2^2).$$

(We are just using the boundary curves of $\mathcal{U}$ as axes for the curvilinear co-ordinates (ρ_1,ρ_2)).

ii) Show that $D = \mathbf{R}^+ = \{d \geq 0\}$ is a cone of variations of at (0,0) and find λ which satisfies theorem 4.7.

iii) Show that for any u in $\mathcal{U}$, $u \neq (0,0)$, there exists a cone of variations D such that the conclusion of theorem 4.7 fails.

4.9 Comments. The proof of theorem 4.7 follows one found in Fleming and Rishel (1975); however inequality constraints are not considered there, so that our proof is a little more complicated. More general multiplier theorems are available, for example in Neustadt (1976), where they are also applied to deterministic optimal control theory. Kushner (1972) was the first to use an abstract multiplier theorem in stochastic control theory, and was then followed by Haussmann (1976).

5 A cone of variations

We shall now use a method, standard in optimal control theory, to construct a cone of variations for the problem (3.15). The idea is to use not small perturbations of u^* on $[0,T]$, but rather arbitrarily large perturbations over short time intervals. We assume that there exist constants k_0, k_1, k_2, K^* such that

(5.1) $|\ell(t,x,u)| + |c(x)| \le k_2\left(1+|x|^q+|u|^q\right)$ for some q in $(0,\infty)$,

(5.2) $dx_t = f(t,x_t,u^*_t)dt + \sigma(t,x_t)dw_t$,

(5.3) $\int_0^T |\sigma(t,x)|^2 dt \le k_0, \quad |f(t,x,u)|^2 \le k_1(1+|x|^2+|u|^2)$

$\sigma(t,x): \mathbf{R}^d \to \mathbf{R}^n$ is onto $F(t,x)$, $|\sigma(t,x)^+| \le k_0$ for all (t,x),

(5.4) $E \exp(\varepsilon|x_0|^2) < \infty$ for some $\varepsilon > 0$,

(5.5) $|u^*(t,x)| \le K^*(1+\|x\|_t)$,

(5.6) $\mathcal{U} \supset \{u$ in $\mathcal{U}_G$: $u(t,\cdot)$ is $\underline{\underline{H}}_t$ measurable$\}$ and $u^*(t,\cdot)$ is $\underline{\underline{H}}_t$ measurable where $\{\underline{\underline{H}}_t\}_{0\le t\le T}$ is a given family of σ-algebras, $\underline{\underline{H}}_t \subset \underline{\underline{G}}^n_t$.

Note that $\{\underline{\underline{H}}_t\}$ need not be increasing! It represents the information available to the controller at time t. We define now $\mathcal{U}^t$, the set of values which the perturbed controls at time t can assume.

Choose $\bar{q} > \max\{2,q\}$ and for each t write $\mathcal{U}^t$ for $L_{\bar{q}}(C^n, \underline{\underline{H}}_t, P\circ x^{-1}; U)$. Since $\underline{\underline{H}}_t \subset \underline{\underline{G}}^n_t$ which is countably generated, then $\mathcal{U}^t$ is separable, cf. Halmos, (1950), pp.168, 177. To obtain separability uniformly with respect to t we make the following stronger hypothesis:

(5.7) (i) there exists a countable set V of measurable functions mapping a measurable space $(A,\underline{\underline{A}})$ into $\left(U,\underline{\underline{B}}(U)\right)$,

and (ii) for each t in $(0,T]$ there exists a measurable mapping i_t mapping $(C^n,\underline{\underline{H}}_t)$ into $(A,\underline{\underline{A}})$, such that for each v in V $v\circ i_t$ is bounded on C^n, and $\{v\circ i_t$: v in V$\}$ is dense in $\mathcal{U}^t$.

Let us see how (5.7) is satisfied in three common examples.

5.1 Example. $\mathcal{U}$ is the set of control laws using complete information. Here $\mathcal{U} = \mathcal{U}_0$ and $\underline{\underline{H}}_t = \underline{\underline{G}}_t^n$. Let β_t be the countable algebra generated by $\{x: x_s \text{ in } B\}$, $s \le t$, s rational, $B \subset \mathbf{R}^n$ a "rectangle" with "rational" vertices. Let $\tilde{U}$ be a countable dense subset of U and let $V = V^T$,

$$V^t = \{ \sum_{i=1}^{N} u_i 1_{A_i}(x): N \text{ finite}, u_i \text{ in } \tilde{U}, A_i \text{ disjoint in } \beta_t \},$$

$$i_t(x) = x^t, \text{ where } x^t(s) = x(t \wedge s).$$

We take for $(A,\underline{\underline{A}})$ the space $(C^n,\underline{\underline{G}}^n)$. Then $i_t: (C^n,\underline{\underline{G}}_t^n) \to (C^n,\underline{\underline{G}}^n)$ is measurable and if v is in V, then $v:(C^n,\underline{\underline{G}}^n) \to \big(U,\underline{\underline{B}}(U)\big)$ is measurable. Clearly $|v \circ i_t(x)| \le \max\{|u_i|: i = 1,2,\cdots,N\}$, hence $v \circ i_t$ is bounded, and V^t is dense in $\mathcal{U}^t$, cf. Halmos (1950). Since $\mathcal{V}^t = \{v \circ i_t: v \text{ in } V^t\}$, and $V^t \subset V$ (since $\beta_t \uparrow \beta_T$), then $\{v \circ i_t: v \text{ in } V\}$ is dense in $\mathcal{U}^t$.

5.2 Example. $\mathcal{U}$ is the set of control laws using partial information. Here $x = \binom{y}{z}$ and $\underline{\underline{H}}_t$ is the σ-algebra generated by $\{y_s: s \le t\}$. Now $\mathcal{U}$ consists of the elements of $\mathcal{U}_0$ which are adapted to $\{\underline{\underline{H}}_t\}$. Let V^t be as above with β_t now generated by $\{x: y_s \text{ in } B\}$, $s \le t$, s rational, B a "rectangle" with "rational" vertices. i_t is again defined as above, and

$$i_t: (C^n,\underline{\underline{H}}_t) \to (C^n,\underline{\underline{H}}_T)$$

is measurable. Note that $(A,\underline{\underline{A}}) = (C^n,\underline{\underline{H}}_T)$ here. Since we still have $\beta_t \uparrow \beta_T$, then as in example 5.1, (5.7) holds.

5.3 Example. $\mathcal{U}$ is the set of Markov controls. Here $\underline{\underline{H}}_t$ is the σ-algebra generated by x_t. Let β be the countable ring generated by "rectangles" in $\mathbf{R}^n$ with "rational" vertices, let

$$V = \{ \sum_{i=1}^{N} u_i 1_{A_i}(y): N \text{ finite}, u_i \text{ in } \tilde{U}, A_i \text{ disjoint in } \beta\}$$

and define $i_t : (C^n, \underline{\underline{H}}_t) \to (\mathbf{R}^n, \underline{\underline{B}}(\mathbf{R}^n))$ by $i_t(x) = x_t$. Then i_t is measurable, $v \circ i_t$ is bounded for v in V, and i_t establishes an isometric isomorphism j,

$$j(v) = v \circ i_t : L_{\bar{q}}(\mathbf{R}^n, \underline{\underline{B}}(\mathbf{R}^n), P \circ x_t^{-1}; U) \to L_{\bar{q}}(C^n, \underline{\underline{H}}_t, P \circ x^{-1}; U)$$

But clearly V is dense in $j^{-1}(\mathcal{U}^t)$, so jV is dense in $\mathcal{U}^t$, and again (5.7) holds.

Let us now construct the promised cone of variations. For v in V define

$$\phi_s^v = \ell(s, x_s, v \circ i_s(x)) - \ell(s, x_s, u^*(s, x_s))$$

and similarly

$$\theta_s^v = \sigma(s, x_s)^+ [f(s, x_s, v \circ i_s(x)) - f(s, x_s, u^*(s, x_s))].$$

Let $\bar{p} = \min\{2\bar{q}(2+\bar{q})^{-1}, \bar{q}\, q^{-1}\}$ and set

$$\psi_s^v = E\{\phi_s^v + \chi_s \theta_s^v\}, \quad \bar{\psi}_s^v = E\{|\phi_s^v + \chi_s \theta_s^v|^{\bar{p}}\}^{1/\bar{p}}.$$

From (5.1), (5.3), (5.4) and the boundedness of v in V, we have

$$|\psi_s^v| \le K(v)[(1+E\|x\|_s^q) + E\{|\chi_s|^2\}^{1/2}(1+E\|x\|_s^2)^{1/2}]$$

$$\bar{\psi}_s^v \le K(v)[(1+E\|x\|_s^{\bar{q}})^{1/\bar{p}} + E\{|\chi_s|^2\}^{1/2}(1+E\|x\|_s^{\bar{q}})^{1/\bar{q}}],$$

for some constant K(v) depending on v. By lemma 3.3 ψ^v, $\bar{\psi}^v$ are integrable functions of s, and there exists a null set T(v) such that for t not in T(v)

$$\frac{d}{dt}\int_0^t \psi_s^v ds = \psi_t^v, \quad \frac{d}{dt}\int_0^t \bar{\psi}_s^v ds = \bar{\psi}_t^v. \tag{5.8}$$

Let $T_0 = \bigcup_{v \in V} T(v)$, so T_0 is a Lebesgue null set. Let D be the convex cone generated by $\{\psi_t^v : t \text{ in } (0,T]\backslash T_0; v \text{ in } V\}$.

5.4 Theorem. D is a cone of variations of $\tilde{J}$ at u^* for problem (3.15).

Proof: Choose d^i in D, $i = 1,2,\cdots,\ell \le m_1+1+m_2$. We may assume that there exist t_j, v_j in V, $j = 1,\cdots,r$, such that $0 < t_1 \le t_2 \le \cdots \le t_r \le T$ and with $\psi_{t_j}^{v_j}$ written as ψ^j

$$d^i = \sum_{j=1}^r a_{ij}\psi^j, \quad a_{ij} \ge 0.$$

We must find $\eta > 0$ and $z: \eta C \to \mathcal{U}\cap\mathcal{U}_G$ such that $z(0) = u^*$, $\tilde{J}\circ z$ is continuous and $M = (d^1, d^2, \cdots, d^\ell)$ is the conical differential of $\tilde{J}$ along z at u^*. For ρ in ηC we have

$$M\rho = \sum_{i=1}^{\ell} d^i \rho_i = \sum_{j=1}^{r} \delta t_j \, \psi^j$$

where $\delta t_j = \sum_{i=1}^{\ell} \rho_i a_{ij} \leq \eta \max_i a_{ij}$. Let s_i be the largest integer such that $t_i = t_j$ if $i \leq j \leq s_i$. Let $\tau_i = \sum_{j=i}^{s_i} \delta t_j$ and let $I_i(\rho)$ be the interval $(t_i - \tau_i, t_i - \tau_i + \delta t_i]$. For η sufficiently small (depending on $\{t_j\}$, $\{a_{ij}\}$) the intervals $I_i(\rho)$ are disjoint subsets of $(0,T]$. Next define

$$(5.9) \qquad u^\rho(t,x) = \begin{cases} v_j \circ i_t(x) \text{ if } t \text{ in } I_j(\rho) \quad j = 1, \cdots, r, \\ u^*(t,x) \text{ if } t \text{ not in } \bigcup_{j=1}^{r} I_j(\rho). \end{cases}$$

We shall complete the proof as a series of lemmas.

5.5 Lemma. If ρ is in ηC for any $\eta < 1$, then u^ρ is in $\mathcal{U}\cap\mathcal{U}_G$.

Proof: (5.6), (5.7) and (5.9) imply that u^ρ is in $\mathcal{U}$ if u^ρ is in $\mathcal{U}_G$, so we need only show the latter. Now (5.5) and the definition of V imply that (3.8) holds for u^ρ, with a constant K_u independent of ρ. Hence u^ρ in $\mathcal{U}_G$. Q.E.D.

Note that from (5.3), (5.5), (3.8) for some constant K_0 independent of ρ

$$(5.10) \qquad |\theta_t^{u^\rho}|^2 \leq 2k_3(2+2|x_t|^2 + |u_t^\rho|^2 + |u_t^*|^2) \leq K_0(1+\|x\|_t^2)$$

$$(5.11) \qquad \int_0^s |\theta_t^{u^\rho}|^2 dt = \sum_{j=1}^{r} \int_{I_j(\rho)\cap[0,s]} |\theta_t^{u^\rho}|^2 dt \leq \xi(1+\|x\|_s^2)$$

where $\xi = K_0 \sum_{j=1}^{r} \delta t_j \leq \eta K_0 \sum_{j=1}^{r} \max_i a_{ij}$.

5.6 Remark. From (5.11) and theorem 2.2 it follows that for any $p < \infty$ there exists η_p such that for $\eta < \eta_p$

$$\sup_{\rho \in \eta C} E|\zeta_T(\theta^{u^\rho})|^p < \infty.$$

We define $z(\rho) = u^\rho$. Then $z: \eta C \to \mathcal{U} \cap \mathcal{U}_G$ and $z(0) = u^*$.

5.7 Lemma. $\tilde{J} \circ z$ is continuous for η sufficiently small.

Proof: We must show that if $\rho_n \to \rho_0$ and if we write u_n for u^{ρ_n}, then $\tilde{J}(u_n) \to \tilde{J}(u_0)$. But

$$\begin{aligned}\tilde{J}(u_n) - \tilde{J}(u_0) &= [E^{u_n} \int_0^T \phi_t^{u_n} dt - E^{u_0} \int_0^T \phi_t^{u_0} dt] \\ &\quad + [E^{u_n} \int_0^T \chi_t \theta_t^{u_n} dt - E^{u_0} \int_0^T \chi_t \theta_t^{u_0} dt].\end{aligned} \tag{5.12}$$

To estimate the second term, write

$$\begin{aligned}&|E^{u_n}\int_0^T \chi_t \theta_t^{u_n} dt - E^{u_0}\int_0^T \chi_t \theta_t^{u_0} dt| \\ &\leq E \int_0^T |\chi_t||\zeta_T(\theta^{u_n})\theta_t^{u_n} - \zeta_T(\theta^{u_0})\theta_t^{u_0}|dt \\ &\leq \{\int_0^T E|\chi_t|^2 dt\}^{1/2}\{E\int_0^T|\zeta_T(\theta^{u_n})\theta_t^{u_n} - \zeta_T(\theta^{u_0})\theta_t^{u_0}|^2 dt\}^{1/2}.\end{aligned}$$

The term inside the last pair of braces is bounded by

$$\begin{aligned}&2E\{\zeta_T(\theta^{u_n})^2\int_0^T|\theta_t^{u_n} - \theta_t^{u_0}|^2 dt\} + 2E\{|\zeta_T(\theta^{u_n}) - \zeta_T(\theta^{u_0})|^2\int_0^T|\theta_t^{u_0}|^2 dt\} \\ &\leq 2E\{\zeta_T(\theta^{u_n})^4\}^{1/2}\, T^{1/2}\{\int_0^T E|\theta_t^{u_n} - \theta_t^{u_0}|^4 dt\}^{1/2} \\ &\quad + 2E\{|\zeta_T(\theta^{u_n}) - \zeta_T(\theta^{u_0})|^4\}^{1/2} T^{1/2}\{\int_0^T E|\theta_t^{u_0}|^4 dt\}^{1/2}.\end{aligned} \tag{5.13}$$

Now (5.10) and lemma 3.3 imply that $|\theta^{u_n}|$ is dominated by a function in

$L_4(dtdP)$, and $|\theta_t^{u_n} - \theta_t^{u_0}| = 0$ except for t in $T_n \equiv \bigcup_j [I_j(\rho_n)\backslash I_j(\rho_0) \cup I_j(\rho_0)\backslash I_j(\rho_n)]$, and the Lebesgue measure of T_n converges to 0 as $n \to \infty$, so by the Lebesgue dominated convergence theorem $\int_0^T E|\theta_t^{u_n} - \theta_t^{u_0}|^4\, dt \to 0$ as $n \to \infty$. This fact and remark 5.6 imply that the first term on the right side of (5.13) converges to 0 for η sufficiently small.

As we observed $\int_0^T E|\theta_t^{u_n} - \theta_t^{u_0}|^2 dt \to 0$ so that $\zeta_T(\theta^{u_n}) \to \zeta_T(\theta^{u_0})$ in probability (cf. exercise 1.5.2). From remark 5.6 again it follows that $|\zeta_T(\theta^{u_n})|^4$ are uniformly integrable if η is small enough so the Lebesgue theorem implies that $E|\zeta_T(\theta^{u_n}) - \zeta(\theta^{u_0})|^4 \to 0$, and so the second term on the right side of (5.13) also converges to 0, i.e. the second term on the right side of (5.12) converges to 0. The first term is treated similarly using (5.1).

Q.E.D.

5.8 <u>Lemma</u>. $\tilde{J}(u^\rho) = \tilde{J}(u^*) + M\rho + o(|\rho|)$.

Proof: From (5.9) and (3.16) it follows that

$$\tilde{J}(u^\rho) - \tilde{J}(u^*) = E^{u^\rho} \sum_{j=1}^r \int_{I_j(\rho)} [\phi_t^{v_j} + \chi_t \theta_t^{v_j}]dt,$$

so we must show that

$$\sum_{j=1}^r E^{u^\rho} \int_{I_j(\rho)} [\phi_t^{v_j} + \chi_t \theta_t^{v_j}]dt = \sum_{j=1}^r E\{\phi_{t_j}^{v_j} + \chi_{t_j} \theta_{t_j}^{v_j}\}\delta t_j + o(|\rho|)$$

i.e. for each j

$$\text{(5.14)} \qquad E\{[\zeta_T(\theta^{u^\rho}) - 1]\int_{I_j(\rho)} (\phi_t^{v_j} - \chi_t \theta_t^{v_j})dt\} + [\int_{I_j(\rho)} \psi_t^{v_j}\, dt - \psi_{t_j}^{v_j}\delta t_j] = o(|\rho|).$$

But

$$\int_{I_j(\rho)} \psi_t^{v_j}\, dt = \int_{t_j-\tau_j}^{t_j-\tau_j+\delta t_j} \psi_t^{v_j}\, dt$$

$$= \int_{t_j-\tau_j}^{t_j} \psi_t^{v_j}\, dt + \int_{t_j}^{t_j-\tau_j+\delta t_j} \psi_t^{v_j}\, dt$$

$$= \tau_j \psi_{t_j}^{v_j} + o(|\tau_j|) + (\delta t_j - \tau_j)\psi_{t_j}^{v_j} + o(|\delta t_j - \tau_j|)$$

by (5.8). Since $\delta t_j = O(|\rho|)$, $\tau_j = O(|\rho|)$ then the second term on the left side of (5.14) is $o(|\rho|)$. Moreover the first term is bounded by $[\bar{p}^{-1}+(\bar{p}')^{-1} = 1]$

$$\int_{I_j(\rho)} E\{|\zeta_T(\theta^{u^\rho}) - 1|^{\bar{p}'}\}^{1/\bar{p}'} \bar{\Phi}_t^{v_j} dt = E\{|\zeta_T(\theta^{u^\rho})-1|^{\bar{p}'}\}^{1/\bar{p}'} O(|\rho|)$$

since again by (5.8)

$$\int_{I_j(\rho)} \bar{\Phi}_t^{v_j} dt = \bar{\Phi}_{t_j}^{v_j} \delta t_j + o(|\rho|) = O(|\rho|).$$

Also

$$\int_0^T E|\theta^{u^\rho}|^2 dt \to 0$$

as $|\rho| \to 0$ cf. (5.11), so that $\zeta_T(\theta^{u^\rho}) \to 1$ in probability. Since $\sup_{\rho \in \eta C} E|\zeta_T(\theta^{u^\rho})|^p < \infty$ for some $p > \bar{p}'$ (if η is small), then $E\{|\zeta_T(\theta^{u^\rho}) - 1|^{\bar{p}'}\}^{1/\bar{p}'} = o(1)$ and so the first term on the left side of (5.14) is also $o(|\rho|)$. Q.E.D.

This completes the proof of the theorem.

5.9 Exercises

5.9.1 Let $U \subset \mathbf{R}^d$ be a Borel set, $\mathcal{U}_0 = \{u: [0,T] \to U$, bounded, measurable$\}$, $A: [0,T) \to \mathbf{R}^n \otimes \mathbf{R}^n$, $\beta: [0,T] \times U \to \mathbf{R}^n$ be bounded, measurable. If $\Phi(t,s)$ is the fundamental matrix solution of

$$\frac{dy}{dt} = A(t)y$$

show that there exists a unique solution of

(5.15)
$$\frac{dx}{dt} = A(t)x + \beta(t,u), \quad x(0) = x_0$$

given by

$$x^u(t) = \Phi(t,0)x_0 + \int_0^t \Phi(t,s)\beta\big(s,u(s)\big)ds.$$

5.9.2 Assume in addition that

$$c: \mathbf{R}^n \to \mathbf{R}^{m_1+1+m_2}$$

is C' (i.e. c and $\partial c/\partial x_i$ are continuous for all i). Let u^*

be in $\mathcal{U}_0$ as defined in 5.9.1.

Define $J:\mathcal{U}_0 \to \mathbf{R}^{m_1+1+m_2}$ by $J(u) = c(x_T^u)$ where $x_t^u = x^u(t)$ as given in 5.9.1. Let V be a countable dense subset of U and for v in V let

$\psi_s^v = \Phi(T,s)[\beta(s,v) - \beta(s,u^*(s))]$. Now let T(v) be the null set on which

$$\frac{d}{dt}\int_0^t \psi_s^v \, ds \neq \psi_t^v$$

and let $T_0 = \bigcup_{v \in V} T(v)$. Show that the convex cone generated by

$$\{c_x(x^{u^*}(T))\psi_t^v : t \text{ in } (0,T]\backslash T_0, \ v \text{ in } V\}$$

is a cone of variations of J at u^*. Here c_x is the matrix $(c_x)_{ij} = \partial c_i/\partial x_j$.

5.10 Comments. The method employed to define perturbed controls is standard in deterministic control theory, cf. Fleming and Rishel (1975). The proof that D is a cone of variations is, of course, particular to this problem. It differs from the proof used by Haussmann (1976), and gives better results. The novel trick is to use (5.7), i.e. the set V and mappings i_t.

6 The abstract necessary conditions

Let us now apply the results of the previous two sections to obtain necessary conditions. Accordingly let u^* solve (3.4), and introduce the notation

(6.1) $$H(t,x,u,p,\lambda) = \lambda\cdot\ell(t,x,u) + p\cdot f(t,x,u)$$

for λ in $\mathbf{R}^{m_1+1+m_2}$, p in $\mathbf{R}^n$. Observe that for u in U

$$\lambda\cdot(\phi_t^u + \chi_t\theta_t^u) = H\big(t,x_t,u,[\chi_t\sigma(t,x_t)^+]'\lambda,\ \lambda\big) - H\big(t,x_t,u^*(t,x),\ [\chi_t\sigma(t,x_t)^+]'\lambda,\lambda\big)$$

where ' denotes transpose.

6.1 Theorem. Assume (5.1) - (5.7), and

(6.2) $f(t,x,\cdot)$, $\ell(t,x,\cdot)$ are continuous for each (t,x).

Then there exists λ in $\mathbf{R}^{m_1+1+m_2}$, $\lambda \neq 0$, $\lambda_i \leq 0$ if $i \geq 0$, $\lambda_i J_i(u^*) = 0$ if $i > 0$, and a Lebesgue null set T_0, such that for each t in $(0,T]\setminus T_0$ and each v in $\mathcal{U}^t$ we have

(6.3) $$E\{H\big(t,x_t,v(x),\tilde{p}_t,\lambda\big)\} \leq E\{H\big(t,x_t,u^*(t,x),\tilde{p}_t,\lambda\big)\}$$

where $\tilde{p}_t' = \lambda'\chi_t\sigma(t,x_t)^+$.

Proof: According to theorems 4.7 and 5.4 the desired λ exists except that (6.3) holds only for v in $\{\bar{v}\circ i_t\colon \bar{v}$ in V$\}$. According to (5.7) for v in $\mathcal{U}^t$ there exists a sequence $\{\bar{v}_m\}$ contained in V such that $\bar{v}_m\circ i_t \to v$ in $L_{\bar{q}}(C^n,\underline{\underline{H}}_t,P\circ x^{-1};U)$, hence a.s. (possibly taking a subsequence). By the definition of $\bar{q}$ there exists $\bar{p} > 1$ (cf. the definition of ψ^v following examples 5.3) such that

$$\sup_m E\left|H\big(t,x_t,\bar{v}_m\circ i_t(x),\tilde{p}_t,\lambda\big)\right|^{\bar{p}} < \infty$$

using (5.1), (5.3), (5.5) and the fact that $\sup_m E|\bar{v}_m\circ i_t(x)|^{\bar{q}} < \infty$ since $E|\bar{v}_m\circ i_t(x)|^{\bar{q}} \to E|v(x)|^{\bar{q}} < \infty$. By Lebesgue's theorem it follows that (6.3) holds for v in $\mathcal{U}^t$. Q.E.D.

6.2 Corollary. Under the assumptions of the theorem, for t in

$(0,T]\backslash T_0$, u in U

(6.4) $$E\{H(t,x_t,u,\tilde{p}_t,\lambda)|\tilde{\underline{\underline{F}}}_t\} \leq E\{H(t,x_t,u^*_t,\tilde{p}_t,\lambda)|\tilde{\underline{\underline{F}}}_t\} \quad \text{a.s.}$$

where $\tilde{\underline{\underline{F}}}_t = x^{-1}(\underline{\underline{H}}_t)$.

Proof: Let B_0 be the set on which (6.4) fails. Then B_0 is in $\tilde{\underline{\underline{F}}}_t$ so $B_0 = x^{-1}(B)$ for some B in $\underline{\underline{H}}_t$. Define

$$v(x) = \begin{cases} u & \text{if } x \text{ in } B \\ u^*(t,x) & \text{otherwise} \end{cases}$$

Then v is in $\mathcal{U}^t$. But for this v (6.3) is violated unless $P\circ x^{-1}(B) = 0$ i.e. $P(B_0) = 0$. Q.E.D.

The above result is not yet satisfactory because we do not have a good representation of χ, i.e. of $\tilde{p}$. Before considering this question, let us establish a converse to the above result; it will prove useful shortly. We assume (5.1) - (5.4) and (6.2).

6.3 Definition. An admissible control u^* is a strong extremal if (5.5), (5.6) hold and if there exists λ in $\mathbf{R}^{m_1+1+m_2}$, $\lambda \neq 0$, $\lambda_i \leq 0$ if $i \geq 0$, $\lambda_i J_i(u^*) = 0$ if $i > 0$, such that for almost all t in (0,T]

(6.5) $$\max_{u\in U} H(t,x_t,u,\tilde{p}_t,\lambda) = H(t,x_t,u^*_t,\tilde{p}_t,\lambda) \quad \text{a.s.}$$

where $\{x_t\}$ corresponds to u^*, and where $\{\tilde{p}_t\}$ is the strong adjoint process defined by

$$\tilde{p}'_t = \lambda'\chi_t\sigma(t,x_t)^+$$

with χ_t given by (3.14), i.e. the martingale representation of $E\{L|\bar{\underline{\underline{F}}}_{t+}\}$ with

$$L = \int_0^T \ell(t,x_t,u^*_t)dt + c(x_T).$$

6.4 Theorem. Assume u^* is a strong extremal and assume that the following constraint qualifications hold:

(a) if $m_1 \neq 0$ then $\{(J_i(u))_{i<0}: u \text{ in } \mathcal{U}_G\}$ contains a neighbourhood of zero in $\mathbf{R}^{m_1}$,

(b) if $m_2 \neq 0$ then there exists u in $\mathcal{U}_G$ such that

$$J_i(u) = 0 \quad \text{if } i < 0$$
$$J_i(u) < 0 \quad \text{if } i > 0 \text{ and } J_i(u^*) = 0.$$

Then u^* solves (3.4) with $\mathcal{U} = \mathcal{U}_G$.

Proof: By denseness (6.5) holds for u replaced by any v in $L_{\bar{q}}(C^n, \underline{\underline{G}}_t^n, P \circ x^{-1}; U)$.

From §3 we know that for u in $\mathcal{U}_G$

$$J(u) = J(u^*) + E^u \int_0^T (\phi_t^u + \chi_t \theta_t^u) dt$$

so that from (6.5), the fact that u_t is in $\mathcal{U}^t$ for u in $\mathcal{U}_G$, and the fact that $P^u \ll P$, it follows that

$$(6.6) \qquad \lambda \cdot J(u) - \lambda \cdot J(u^*) = \int_0^T E^u \lambda \cdot (\phi_t^u + \chi_t \theta_t^u) dt \leq 0.$$

Now if $\lambda_0 = 0$ then (b) allows us to choose u in $\mathcal{U}_G$ for which (6.6) implies that $\lambda_i = 0$ for $i > 0$, and similarly (a) implies that $\lambda_i = 0$ for $i < 0$, contradicting $\lambda \neq 0$. Thus the constraint qualifications eliminate the abnormal case $\lambda_0 = 0$.

If u is any candidate for the inf, i.e. if u satisfies the constraints, then (6.6) gives

$$\lambda_0 J_0(u) \leq \lambda \cdot J(u) \leq \lambda \cdot J(u^*) = \lambda_0 J_0(u^*)$$

and hence $J_0(u) \geq J_0(u^*)$ since $\lambda_0 < 0$. Q.E.D.

6.5 Corollary. Assume $m_1 = m_2 = 0$. Assume (5.1) - (5.5) and (6.2) as well as $\underline{\underline{F}}_t = \underline{\underline{F}}_t^x$. Then u^* solves

$$\inf\{J_0(u): u \text{ in } \mathcal{U}_G\}$$

if and only if it is a strong extremal.

Proof: The necessity follows from corollary 6.2 since (5.6) is satisfied with $\underline{\underline{H}}_t = \underline{\underline{G}}_t^n$, $\mathcal{U} = \mathcal{U}_G$, since $\tilde{\underline{\underline{F}}}_t = x^{-1}(\underline{\underline{G}}_t^n) = \underline{\underline{F}}_t^x = \underline{\underline{F}}_t$, and since H is measurable with respect to $\underline{\underline{F}}_t$. The sufficiency follows from theorem 6.4 since $m_1 = m_2 = 0$. Q.E.D.

6.6 Remark. The condition $\underline{\underline{F}}_t = \underline{\underline{F}}_t^x$ can be dispensed with because w can be taken to be a Brownian motion with respect to $\{\underline{\underline{F}}_t^x\}$ so that $\underline{\underline{F}}_t$ can be replaced by $\underline{\underline{F}}_t^x$, cf. Wong (1971).

Hence for the problem $\inf\{J_0(u): u \text{ in } \mathcal{U}_G\}$ the conditions of the abstract maximum principle (corollary 6.2) are both necessary and sufficient.

6.7 Corollary. Assume (5.1) - (5.6) and (6.2) with $\underline{\underline{H}}_t = \underline{\underline{G}}_t^n$. If u^* solves (3.4) then there exists λ in $\mathbf{R}^{m_1+1+m_2}$ such that u^* solves

(6.7) $$\inf\{\bar{J}(u) : u \text{ in } \mathcal{U}_G\}$$

where $\bar{J}(u) = E\{\int_0^T - \lambda \cdot \ell(t,x_t,u_t)dt - \lambda \cdot c(x_T)\}$.

Proof: (5.7) is satisfied as in example 5.1. Let λ be given by theorem 6.1. Then for almost all t and for each u in U,

(6.8) $$H(t,x_t,u,\tilde{p}_t,\lambda) \leq H(t,x_t,u_t^*,\tilde{p}_t,\lambda) \quad \text{a.s.}$$

But by corollary 6.5 u^* solves (6.7) if and only if

(6.9) $$\bar{H}(t,x_t,u,\bar{p}_t,-1) \leq \bar{H}(t,x_t,u_t^*,\bar{p}_t,-1) \quad \text{a.s.}$$

where $\bar{H}(t,x,u,p,-1) = -1[-\lambda \cdot \ell(t,x,u)] + p \cdot f(t,x,u)$ and $\bar{p}_t' = \bar{\chi}_t \sigma(t,x_t)^+$ and [cf. (3.14)]

$$\int_0^T -\lambda \cdot \ell(t,x_t,u_t^*)dt - \lambda \cdot c(x_t)$$

$$= E\{\int_0^T -\lambda \cdot \ell dt - \lambda \cdot c | \underline{\underline{F}}_0\} + \int_0^T \bar{\chi}_s dw_s + \bar{M}_t.$$

Comparing with (3.14) we see that $-\lambda'\chi_t = \bar{\chi}_t$, $-\lambda' M_t = \bar{M}_t$. Hence $\bar{p}_t = \tilde{p}_t$ and $\bar{H}(t,x_t,u,\bar{p}_t,-1) = H(t,x_t,u,\tilde{p}_t,\lambda)$. Now (6.8) implies (6.9) and so u^* solves (6.7). Q.E.D.

By finding a problem equivalent to (6.7) for which we can find a necessary condition readily, and then by translating this condition back to the original problem (3.4), we shall obtain the desired maximum principle. This program constitutes the remainder of the proof.

6.8 Exercise

6.8.1 We continue the exercises 5.9.1, 5.9.2. Assume in addition that $u \to \beta(t,u)$ is continuous for almost all t. If u^* is a solution of

$$\inf\{J_0(u) : u \text{ in } \mathcal{U}\}$$

show that there exists λ in $R^{m_1+1+m_2}$, $\lambda \neq 0$, $\lambda_i \leq 0$ if $i \geq 0$, and $\lambda_i J_i(u^*) = 0$ if $i \neq 0$, as well as an absolutely continuous function $p:[0,T] \to \mathbf{R}^n$ which satisfies

$$\frac{dp}{dt} = -A(t)'p, \quad p(T) = c_x\left(x^{u^*}(T)\right)'\lambda,$$

such that for almost all t and all u in U

$$p(t)\cdot[A(t)x^{u^*}(t) + \beta(t,u)] \leq p(t)\cdot[A(t)x^{u^*}(t) + \beta(t,u^*(t))].$$

6.9 Comments. Corollary 6.2 first appeared in Haussmann (1976). In the Markov case with complete observation (example 5.3) one can identify the strong adjoint process with the negative of the gradient of the value function, as was done in Haussmann (1981), but this approach links the maximum principle to the Hamilton-Jacobi-Bellman equation which we prefer not to do. This link explains why the necessary condition is also sufficient, and indeed the results of this section are closely related to the work of Bismut (1976) and of Davis (1973).

7 An equivalent problem

In principle corollary 6.2 constitutes a maximum principle, but because the adjoint process $\tilde{p}$ contains the unknown χ, this result is not very useful; see however §11 where the sufficiency results of corollary 6.5 are used. On the other hand, in the case of complete observation we can use the Lagrange multiplier feature of these conditions to define a problem without constraints which is solved by any solution of the original problem (3.15), cf. corollary 6.7. We shall now show that for such problems without constraints we can enlarge the set of admissible controls to be all adapted stochastic processes. This result is in the spirit of corollary 4.2 in chapter VI of Fleming and Rishel (1975).

Let us motivate why controls which are adapted stochastic processes are preferable to feedback controls in the calculation of necessary conditions. The basis for a necessary condition (in the absence of constraints) is the inequality

$$J_0(u) - J_0(u^*) \geq 0$$

or $\delta J_0(u^*; u-u^*) \geq 0$. However to compute this variation of J_0, we need to compute δx_t^*, the perturbation of x_t^*. If $u_t^* = u^*(t,\omega)$, then the perturbation δx_t^* at time t due to a perturbation in u^* prior to t generates as

$$d(\delta x_t^*) = f_x(t,x_t^*,u_t^*)(\delta x_t^*)dt + \sigma_x(t,x_t^*)(\delta x_t^*)dw_t,$$

whereas if $u^* = u^*(t,x_t^*)$, then this equation would be

$$d(\delta x_t^*) = [f_x(t,x_t^*,u_t^*) + f_u(t,x_t^*,u_t^*)\frac{\partial u_t^*}{\partial x}](\delta x_t^*)dt$$
$$+ \sigma_x(t,x_t^*)(\delta x_t^*)dw_t.$$

Since $\frac{\partial u^*}{\partial x}$ may not exist, this latter situation must be avoided, hence our effort to switch to u^* in the form $u^*(t,\omega)$.

Suppose that u^* solves the completely observable problem without constraints:

(7.1) $$\inf\{J_0(u): u \text{ in } \mathcal{U}_G\},$$

so that corresponding to $u^*(t,x)$ we have $(\Omega,\underline{\underline{F}},\{\underline{\underline{F}}_t\},P,\{w_t\},\{x_t^*\})$ where $\{x_t^*\}$ satisfies

(7.2) $$dx_t = f(t,x_t,u_t^*)dt + \sigma(t,x_t)dw_t,$$

with $u_t^* = u^*(t,x^*)$. The problem in corollary 6.7 has this form. Since we shall be talking about strong solutions, we use * to denote the solution corresponding to u^*. Recall that for u in $\mathcal{U}_G$, the corresponding solution of (7.2) is defined using Girsanov's theorem, hence is a weak solution (assuming (7.3),(7.5),(7.6) and (7.8) below). What we shall now do is to find another problem solved by u^* for which the solutions of (7.2) are strong.

Let $\bar{q} > \max\{2,q\}$, and let $\tilde{\mathcal{U}}$ be the set of measurable adapted processes on $(\Omega,\underline{\underline{F}},\{\underline{\underline{F}}_t\},P)$ which are in $L_{\bar{q}}([0,T]\times\Omega;U)$. Note that $\{u_t^*(\omega)\} = \{u^*(t,x^*(\omega))\}$ is in $\tilde{\mathcal{U}}$ assuming (7.3), (7.6) and (7.8) from the following list:

(7.3) $|f(t,x,u)|^2 \leq k_1(1+|x|^2+|u|^2)$,

(7.4) $|f(t,x,u) - f(t,y,u)| + |\sigma(t,x) - \sigma(t,y)| \leq k_1|x-y|$,

(7.5) $\sigma(t,x)\colon \mathbf{R}^d \to \mathbf{R}^n$ is onto $F(t,x)$ and $|\sigma(t,x)^+| \leq k_0$,

$$\int_0^T |\sigma(t,x)|^2 dt \leq k_0,$$

(7.6) $|u^*(t,x)| \leq K^*(1+\|x\|_t)$,

(7.7) $|\ell(t,x,u)| + |c(x)| \leq k_2(1+|x|^q+|u|^q)$,

(7.8) $E\exp\{\varepsilon|x_0|^2\} = \int e^{\varepsilon|x|^2}\mu(dx) < \infty$ for some $\varepsilon > 0$,

(7.9) f is continuous in u for each (t,x),

(7.10) c is continuous, ℓ is continuous in (x,u) for each t

Note that (7.4) and (7.9) imply that f is continuous in (x,u) for each t.

We will show that u^* also solves

(7.11) $\inf\{J_0(u)\colon u \text{ in } \tilde{\mathcal{U}}\}$.

Let us point out that here we have dropped the subscript 0 on ℓ_0, c_0, so that

$$J_0(u) = E\{\int_0^T \ell(t,x_t^u,u_t)dt + c(x_T^u)\}$$

and x^u is the strong (Itô) solution of (7.2) which exists if we assume (7.3), (7.4). Moreover $J_0(u)$ is finite for u in $\tilde{\mathcal{U}}$ assuming (7.7), (7.8). In fact we have

7.1 Lemma. Assume (7.3)-(7.5) and let x^{yu} denote the strong solution of (7.2) for u in $\tilde{\mathcal{U}}$ with $x_0^{yu} = y$. For any constants c_0, c_1, and $p \geq 2$, there exists a constant c_2 such that

$$\sup\{E|x_t^{yu}|^p: 0 \leq t \leq T, \quad E|y|^p \leq c_0, \int_0^T E|u_t|^p dt \leq c_1\} \leq c_2.$$

Proof: Apply Gronwall's lemma to $E|x_t|^p$, using (7.3) and theorem 3.8.2.

Q.E.D.

The following is a useful continuity result.

7.2 Lemma. Assume (7.3)-(7.5) and (7.9). If for almost all t $u_t^n \to u_t^0$ in probability, if $\sup_n \int_0^T E|u_t^n|^{\bar{q}} dt < \infty$, and if $E|y^n - y^0|^2 \to 0$, then $E\|x^{y_n,u_n} - x^{y_0,u_0}\|_T^2 \to 0$.

Proof: We write x^n for x^{y_n,u_n}, $n = 0,1,2,\cdots$. Then from (7.4)

$$\begin{aligned}
E\|x^n - x^0\|_t^2 \leq\ & 3\{E|y^n - y^0|^2 + 2T\int_0^t E|f(s,x_s^0,u_s^n) - f(s,x_s^0,u_s^0)|^2 ds \\
& + 2T\int_0^t E\{|f(s,x_s^n,u_s^n) - f(s,x_s^0,u_s^n)|^2\}ds \\
& + E\|\int_0^{\bullet} \sigma(s,x_s^n) - \sigma(s,x_s^0) dw_s\|_t^2\} \\
\leq\ & K\{E|y^n - y^0|^2 + \int_0^T E|f(s,x_s^0,u_s^n) - f(s,x_s^0,u_s^0)|^2 ds \\
& + \int_0^t E|x_s^n - x_s^0|^2 ds\}
\end{aligned}$$

using the martingale inequality or theorem 3.8.2. Now Gronwall's inequality gives (since $|x_s^n - x_s^0| \leq \|x^n - x^0\|_s$)

$$E\|x^n - x\|_T^2 \leq \bar{K}\{E|y^n - y^0|^2 + \int_0^T E|f(s,x_s^0,u_s^n) - f(s,x_s^0,u_s^0)|^2 ds\}.$$

It remains only to show that the integral on the right side converges to 0.

By lemma 7.1 $E|x_t^0|^2 < \infty$ for each t, so for each t, η, there exists N such that

$$\Pr\{|x_t^0| > N\} \leq \frac{1}{N^2} E|x_t^0|^2 < \eta/4.$$

Moreover for almost all t, $E|u_t^0|^q$ is finite so

$$P\{\{|u_t^n| > M\} \cup \{|u_t^0| > M\}\} \leq P\{|u_t^n - u_t^0| > 1\} + P\{|u_t^0| > M - 1\}$$

$$\leq \eta/4 + E|u_t^0|^{\bar{q}}/(M-1)^{\bar{q}} < \eta/2$$

if n and M are sufficiently large. Since $f(t,\cdot,\cdot)$ is uniformly continuous on the compact set $\{|x| \leq N\}\times\{|u| \leq M\}$, then it is continuous in u uniformly in x; hence given $\varepsilon > 0$ there exists $\delta > 0$ such that

$$\sup_{|x|\leq N} |f(t,x,u) - f(t,x,v)| \leq \varepsilon \quad \text{if } |u-v| \leq \delta \quad \text{and} \quad |u| \leq M,\ |v| \leq M.$$

Thus for n sufficiently large

$$P\{|f(t,x_t^0,u_t^n) - f(t,x_t^0,u_t^0)| > \varepsilon\}$$

$$\leq P\{|x_t^0| > N\} + P\{|u_t^n|\ |u_t^0| > M\} + P\{|u_t^n-u_t^0| > \delta\} < \eta.$$

It follows that $f(s,x_s^0,u_s^n) \to f(s,x_s^0,u_s^0)$ in probability a.e.t., hence in measure (dtdP) since $T < \infty$ and we can apply the theorems of Fubini and Lebesgue. But $\bar{q} > 2$ and (7.3) implies

$$\sup_n \int_0^T E|f(s,x_s^0,u_s^n)|^{\bar{q}}ds \leq \sup_n K \int_0^T(1+E|x_s^0|^{\bar{q}} + E|u_s^n|^{\bar{q}})ds < \infty$$

so the result follows by uniform integrability and Lebesgue's theorem.

Q.E.D.

7.3 Lemma. Assume (7.3)-(7.5), (7.7) - (7.9). If $u^n \to u^0$ ae(dtdP) and if $|u_t^n| \leq \nu_t$ where $\int_0^T E|\nu_t|^{\bar{q}}dt < \infty$, then $J_0(u^n) \to J_0(u^0)$.

Proof: We will show that any subsequence of $\{J_0(u^n)\}$ contains itself a subsequence which converges to $J_0(u^0)$. Let $\{J_0(u^n)\}$ be a subsequence.

By lemma 7.2 $E\|x^{x_0,u^n} - x^{x_0,u^0}\|_T^2 \to 0$, i.e. for a subsequence $x_t^n \to x_t^0$ a.s., all t (we write $x^n = x^{x_0,u^n}$). By continuity $\ell(t,x_t^n,u_t^n) \to \ell(t,x_t^0,u_t^0)$ a.e. (dtdP). Since $\bar{q} > q$ and $\sup_n \int_0^T E|\ell(t,x_t^n,u_t^n)|^{\bar{q}/q}dt \leq \sup_n K \int_0^T(1+E|x_t^n|^{\bar{q}} + E|u_t^n|^{\bar{q}})dt < \infty$ by lemma 7.1 and (7.7), then by uniform integrability the convergence of ℓ is also in L_1(dtdP). Similarly for $c(x_T^n)$.

Q.E.D.

We can now proceed with the main result.

7.4 Theorem. Assume (7.3) - (7.10). If U is closed, then

$$\inf\{J_0(u): u \text{ in } \mathcal{U}_G\} \leq \inf\{J_0(u): u \text{ in } \tilde{\mathcal{U}}\} \tag{7.12}$$

Proof: It suffices to show that for any u in $\tilde{\mathcal{U}}$, $\varepsilon > 0$, there exists $\bar{u}$ in $\mathcal{U}_G$ such that $J_0(\bar{u}) \leq J_0(u) + \varepsilon$. We first reduce to the case where u is bounded. Let $1_N(u)$ be the indicator function of $\{u: |u| \leq N\}$ and define

$$u_t^N = 1_N(u_t)u_t + [1 - 1_N(u_t)]u_0$$

for some u_0 in U. Since $1_N(u_t) \uparrow 1$ a.e. (dtdP), then $u^N \to u$ a.e. (dtdP), and $|u_t^N| \leq |u_t| + |u_0|$ is in $L_{\bar{q}}$(dtdP). By lemma 7.3 $J_0(u^N) \to J_0(u)$, and of course u^N is in $\tilde{\mathcal{U}}$, $|u_t^N(\omega)| \leq N + |u_0|$. Hence we may take u to be bounded, or equivalently we may replace U by the compact set $U \cap \{u: |u| \leq N + |u_0|\}$. From now on we assume U to be compact.

For each k there exists a partition $0 = t_0 < t_1 < \cdots < t_k = T$ such that if $u_t^k = u_{t_i}$ for $t_i < t \leq t_{i+1}$, $u_0^k = u_0$, then $u^k \to u$ a.e. (dtdP) as $k \to \infty$ cf. lemma 7.5.1. Note that there are other step function approximations of u (for example $(t_i - t_{i-1})^{-1} \int_{t_{i-1}}^{t_i} u_s ds$ for $t_i < t \leq t_{i+1}$) but they are not necessarily U-valued unless U is convex. Since U is compact then u^k is bounded uniformly in k, so by lemma 7.3 again, $J_0(u^k) \to J_0(u)$ and we need only show that for such a step-function u^k we have a $\bar{u}$ in $\mathcal{U}_G$ such that $J_0(\bar{u}) \leq J_0(u^k) + \varepsilon$.

We fix the partition $0 = t_0 < t_1 < \cdots < t_k = T$. Let $\tilde{\mathcal{U}}^k$ be the set of U-valued processes such that on $(t_i, t_{i+1}]$ the process equals an $\mathcal{F}_{t_i}$-measurable random variable. If i = 0 use $[t_0, t_1]$. Let $\mathcal{U}^k$ be the set of elements of $\mathcal{U}_G$ such that on $(t_i, t_{i+1}]$, $u(t,x) = u_i(x)$, with u_i a $\mathcal{B}^n_{t_i}$ measurable function. Observe that u in $\tilde{\mathcal{U}}^k$ can be thought of as a U^k-valued function on Ω, and u in $\mathcal{U}^k$ as a U^k-valued function on C^n. For u in $\tilde{\mathcal{U}}^k$ we write x^u for the corresponding strong solution which exists because (7.3), (7.4) hold. For $\bar{u}$ in $\mathcal{U}^k$ we can solve (7.2) successively on each segment $(t_i, t_{i+1}]$, since on such a segment

$$f(t, x_t, \bar{u}(t,x)) = f(t, x_t, \bar{u}_i(x))$$

and $\bar{u}_i(x)$ depends only on the past of x up to t_i, i.e. $\bar{u}_i(x) = \bar{u}_i(x^{\bar{u}}(\omega))$ is $\underline{\underline{F}}_{t_i}$-measurable. Hence Itô's theorem implies that a unique strong solution exists on $[t_i, t_{i+1}]$, thus on $[0,T]$. By law uniqueness (since we have pathwise uniqueness) $\{x_t^{\bar{u}}\}$ under P has the same distribution as $\{x_t^*\}$ under $P^{\bar{u}}$, so that

$$J_0(\bar{u}) = E^{\bar{u}}\{\int_0^T \ell(t,x_t^*,\bar{u}(t,x_t^*))dt + c(x_t^*)\}$$

$$= E\{\int_0^T \ell(t,x_t^{\bar{u}},\bar{u}(t,x^{\bar{u}}))dt + c(x_T^{\bar{u}})\}.$$

We shall use the second evaluation because now $x^{\bar{u}}$ is a strong solution on the same $(\Omega,\underline{\underline{F}},\{\underline{\underline{F}}_t\},P,\{w_t\})$ for all $\bar{u}$.

One further reduction will ease the subsequent proof. For v in U^k, x in C^n we let

$$\bar{\phi}(x,v) = \int_0^T \ell(t,x_t, \sum_{i=1}^k v_i 1_{(t_{i-1},t_i]}(t))dt + c(x_T),$$

$$\phi(x,v) = \begin{cases} \bar{\phi}(x,v) & \text{if } |\bar{\phi}(x,v)| \leq M \\ \dfrac{\bar{\phi}(x,v)M}{|\bar{\phi}(x,v)|} & \text{otherwise.} \end{cases}$$

Then ϕ is bounded by M and is continuous on $C^n \times U^k$ by (7.10), (7.7) and Lebesgue's theorem. Moreover for v in $\mathcal{U}^k$ or $\tilde{\mathcal{U}}^k$ it follows from (7.7) and lemma 7.1 (n.b. U is bounded so we may drop $|u|^q$ from (7.7)) that

$$|J_0(v) - E\ \phi(x^v(\omega),v(\omega))| \leq E|\bar{\phi}(x^v,v) - \phi(x^v,v)| \to 0$$

as $M \to \infty$, because $|\phi(x^v,v)|\uparrow|\bar{\phi}(x^v,v)|$ a.s., and $E|\bar{\phi}(x^v,v)| < \infty$.

To complete the proof we will show that for any ϕ, bounded by M and continuous on $C^n \times U^k$, and for any u in $\tilde{\mathcal{U}}^k$, $\varepsilon > 0$, there exists $\bar{u}$ in $\mathcal{U}^k$ such that

(7.13) $$E\ \phi(x^{\bar{u}}, \bar{u}(x^{\bar{u}})) \leq E\ \phi(x^u,u) + \varepsilon$$

Note the U is taken as compact. The proof is by induction on k. For k = 1 let $\gamma(y,v) = E\ \phi(x^{yv},v)$, v in U, y in $\mathbf{R}^n$. γ is continuous because if $(y^n, v^n) \to (y, v)$ then by lemma 7.2, $x^n(\omega) \equiv x^{y^n, v^n}(\omega) \to x^{y,v}(\omega)$ in C^n in probability so $\psi(x^n, v^n) \to \psi(x^{y,v}, v)$ in probability, i.e. $\gamma(y^n, v^n) \to \gamma(y, v)$ since ψ is bounded. Hence there exists a U-valued function $v(y)$ such that

$$\beta(y) \equiv \inf\{\gamma(y,v): v \text{ in } U\} = \gamma\big(y, v(y)\big).$$

Observe that β is continuous for if $y_n \to y$ then by uniform continuity of γ on compact subsets, for any $\varepsilon > 0$,

$$\beta(y) \leq \gamma\big(y, v(y_n)\big) \leq \gamma\big(y_n, v(y_n)\big) + \varepsilon = \beta(y_n) + \varepsilon$$

$$\leq \gamma\big(y_n, v(y)\big) + \varepsilon \leq \gamma\big(y, v(y)\big) + 2\varepsilon = \beta(y) + 2\varepsilon$$

if n is sufficiently large. It follows that v can be chosen to be Borel measurable (cf. theorem 7.5.2). If we define $\bar{u}(t,x) = v(x_0)$ then $\bar{u}$ is in $\mathcal{U}^1$ and

$$E\,\psi(x^v, v) = E\,E\{\psi(x^v, v)|\underline{\underline{F}}_0\} = E\,\gamma(x_0, v) \geq E\,\gamma(x_0, \bar{u}) = E\,\psi\big(x^{\bar{u}}, \bar{u}(x^{\bar{u}})\big).$$

This establishes (7.13) for k = 1.

Now assume that the result holds for k. Take ν in U^k and v in U, $\bar{y}$ in C^n and write y for $\bar{y}(t_k)$. Let x^{yv} be the solution of (7.2) on $(t_k, t_{k+1}]$ with $x^{y,v}_{t_k} = y$ and with $u_t \equiv v$. Define

$$\gamma(\bar{y}, \nu, v) = E\{\psi[(\bar{y}, x^{y,v}), (\nu, v)]\}$$

$$\beta(\bar{y}, \nu) = \inf\{\gamma(\bar{y}, \nu, v): v \text{ in } U\}.$$

As for k = 1, by lemma 7.2, γ is continuous and bounded, β is continuous and bounded, and β is actually the minimum, i.e. the infinum is attained. Again by measurable selection, there is a Borel measurable v such that $\beta(\bar{y}, \nu) = \gamma\big(\bar{y}, \nu, v(\bar{y}, \nu)\big)$.

Let u be in $\tilde{\mathcal{U}}^{k+1}$, so $u = (u^k, u^0)$ where u^k is in $\tilde{\mathcal{U}}^k$ and u^0 is U-valued, $\underline{\underline{F}}_{t_k}$-measurable. Let x^{u^k} be the corresponding solution on $[0, t_k]$ and write y for $x^{u^k}_{t_k}$. By the induction hypothesis there exists $\bar{u}^k$ in $\mathcal{U}^k$ such that $E\,\beta\big(x^{\bar{u}^k}, \bar{u}^k(x^{\bar{u}^k})\big) \leq E\,\beta(x^{u^k}, u^k) + \varepsilon$. Let $\bar{u} = \big(\bar{u}^k, v(x^{\bar{u}^k}, \bar{u}^k(x^{\bar{u}^k}))\big)$. Then $\bar{u}$ is in $\mathcal{U}^{k+1}$ because on $(t_k, t_{k+1}]$ it has the value $v\big(x^k, \bar{u}^k(x^k)\big) = v_{k+1}(x^0)$ (if x in $C^n[0, t_{k+1}]$ is written as (x^k, x^0) with x^k in $C^n[0, t_k]$) and v_{K+1} is $\underline{\underline{G}}^n_{t_k}$-measurable. Indeed

$x \to \bar{u}^k$ U^k is $\underline{\underline{G}}^n_{t_k}$ measurable, and hence, since v is measurable $\underline{\underline{G}}^n_{t_k} \otimes \underline{\underline{B}}(U^k) \to \underline{\underline{B}}(U)$, then v_{k+1} is measurable $\underline{\underline{G}}^n_{t_k} \to \underline{\underline{B}}(U)$. Thus

$$
\begin{aligned}
E\,\psi\big(x^{\bar{u}},\bar{u}(x^{\bar{u}})\big) &= E\,E\{\psi[(x^{\bar{u}^k}, x^{y,v_{k+1}}),(\bar{u}^k,v_{k+1})]\,|\,\underline{\underline{F}}_{t_k}\} \\
&= E\,\gamma\big(x^{\bar{u}^k},\bar{u}^k,v\big(x^{\bar{u}^k},\bar{u}^k(x^{\bar{u}^k})\big)\big) \\
&= E\,\beta\big(x^{\bar{u}^k},\bar{u}^k(x^{\bar{u}^k})\big) \\
&\leq E\,\beta(x^{u^k},u^k) + \varepsilon \\
&\leq E\,\gamma(x^{u^k},u^k,u^0) + \varepsilon \\
&= E\,E\{\psi[(x^{u^k},x^{y,u^0}),(u^k,u^0)]\,|\,\underline{\underline{F}}_{t_k}\} + \varepsilon \\
&= E\,\psi(x^{u^k},u^k) + \varepsilon.
\end{aligned}
$$

This establishes (7.13) for k+1. Q.E.D.

Since u^* is in $\tilde{\mathcal{U}}$ then theorem 7.4 implies that u^* solves

(7.14) $\inf\{J_0(u): u \text{ in } \tilde{\mathcal{U}}\}$.

7.5 Appendix

7.5.1 Lemma. Given $b(t,\omega)$ such that $E\int_0^T|b(t,\omega)|^p dt < \infty$ for some $p > 1$, then there exists $\tilde{s}$ in $[0,T]$ such that $g_n \to b$ a.e. (dtdP) if

$$g_n(t,\omega) = b(\tilde{s}+j2^{-n},\omega) \text{ for } \tilde{s} + j2^{-n} < t \leq \tilde{s} + (j+1)2^{-n} \quad j=0,\pm1,\pm2,\cdots$$

Proof: We follow Liptser and Shiryayev (1977). Without loss of generality we may take $|b(t,\omega)| < \infty$ everywhere by redefining it on a negligible set. Extend b by

$$b(t,\omega) = \begin{cases} b(0,\omega) & \text{if } -T-1 \leq t \leq 0, \\ 0 & \text{if } t < -T-1 \text{ or } t > T. \end{cases}$$

Let

$$b_m(t,\omega) = m\int_{t-\frac{1}{m}}^{t} b(s,\omega)ds.$$

Then b_m is continuous in t and

$$\lim_{m\to\infty} b_m(t,\omega) = \frac{d}{dt}\int_0^t b(s,\omega)ds = b(t,\omega) \quad \text{a.e. (dtdP)}.$$

Also since b = 0 outside [-T-1,T] then

$$E \int_{-T-1}^{2T} |b_m|^p dt \leq E \int_{-T-1}^{2T} m \int_{t-\frac{1}{m}}^{t} |b(s)|^p ds dt$$

$$\leq E \int_{-T-1}^{2T} |b(s)|^p ds < \infty.$$

Lebesgue's theorem implies that

$$E \int_{-T-1}^{2T} |b_m - b| ds \to 0$$

and hence

$$E \int_0^T |b_m(s+t+h) - b(s+t+h)| ds \to 0$$

uniformly in (t,h) in $[-T,T]\times[-1,0]$. Using the triangle inequality, the Minkowski inequality and the continuity of b_m, it follows that

$$E \int_0^T |b(s+t+h) - b(s+t)| ds \to 0$$

as $h \uparrow 0$ if $|t| \leq T$.

Let

$$\psi_n(t) = j2^{-n} \quad \text{if} \quad j2^{-n} < t \leq (j+1)2^{-n}, \quad j=0,\pm 1,\pm 2,\cdots$$

Then for $|t| \leq T$,

$$\lim_{n\to\infty} E \int_0^T |b(s+\psi_n(t)) - b(s+t)| ds = 0$$

Since

$$E \int_0^T |b(s+\psi_n(t))|^p ds \leq E \int_{t-2^{-n}}^{T+t} |b|^p ds < \infty$$

then

$$\lim_{n\to\infty} E \int_{-T}^{T} \int_0^T |b(s+\psi_n(t)) - b(s+t)| ds dt = 0,$$

and hence there exists a subsequence such that for $|t| \leq T$, $0 \leq s \leq T$

$$b(s+\psi_n(t),\omega) \to b(s+t,\omega) \qquad \text{a.e. } (dtdsdP)$$

Let v = s+t. Then there exists $\tilde{s}$ in [0,T] such that

$$b(\tilde{s}+\psi_n(v-\tilde{s}),\omega) \to b(v,\omega) \qquad \text{a.e. } (dvdP)$$

In fact this convergence holds for all but a negligible set of $\tilde{s}$. The result follows with

$$g_n(t,\omega) = b(\tilde{s}+\psi_n(t-\tilde{s}),\omega). \qquad \text{Q.E.D.}$$

7.5.2 Theorem (measurable selection). Assume X is a Banach space, U is a compact subset of a Euclidean space, $\beta: X \to R$ and $\gamma: X \times U \to R$ are continuous, and $\beta(y)$ is in $\gamma(y,U)$ for all y in X. Then there exists a Borel measurable function $u: X \to U$, such that $\beta(y) = \gamma(y,u(y))$.

Proof: U is compact, hence there exists a finite disjoint partition $\{V_j^0\}_{j=1}^N$ of U with diam $V_j^0 \leq 1$. Let d be the dimension of the space containing U. Each set V_j^0 can be partitioned into 2^d pieces of diameter at most $2^{-1/2}$ [each V_j^0 lies in a sphere of radius 1: the new partition consists of the pieces in the various orthants]. This forms a new partition of U, $\{V_j^1\}_{j=1}^{N2^d}$, with $V_i^1 \subset V_j^0$ for $2^d(j-1) < i \leq 2^d j$, $V_j^0 = \bigcup_{2^d(j-1)+1}^{2^d j} V_i^1$. Continue in this manner to refine each partition to obtain $\{V_j^n\}_{j=1}^{N2^{nd}}$, diam $V_j^n \leq 2^{-n/2}$, and $V_j^{n-1} = \bigcup_{2^d(j-1)+1}^{2^d j} V_i^n$. Let $U_j^n = \bigcup_{i \leq j} \overline{V_i^n}$, and let

$$B_j^n = \{y: \beta(y) \text{ in } \gamma(y,U_j^n) \backslash \gamma(y,U_{j-1}^n)\}$$

$$= \{y: \beta(y) \text{ in } \gamma(y,U_j^n)\} \backslash \{y: \beta(y) \text{ in } \gamma(y,U_{j-1}^n)\}.$$

Note that $\{B_j^n\}_{j=1}^{N2^{nd}}$ is a disjoint partition of X. Since U_j^n is compact and β, γ are continuous, then it can readily be shown that $\{y: \beta(y) \text{ in } \gamma(y,U_j^n)\}$ is closed, hence B_j^n is a Borel set. Choose u_j^n in V_j^n and let

$$u^n(y) = \sum_{j=1}^{N2^{nd}} u_j^n \, 1_{B_j^n}(y).$$

Then u^n is U-valued and Borel measurable.

Let $j^n(x) = j$ for x in B_j^n. Since $U_j^n = U_{2^d j}^{n+1}$, then $j^{n+1}(x)$ belongs to $\{2^d j^n(x) - 2^d + 1, \cdots, 2^d j^n(x)\}$, and hence $u^{n+1}(y) = u_{j^{n+1}}^{n+1}(y)$

s in $\bigcup_{i=2^d j^n(y)-2^d+1}^{2^d j^n(y)} V_i^{n+1} = V^n_{j^n(y)}$, and

$$|u^n(y)-u^{n+1}(y)| \leq \text{diam } V^n_{j^n(y)} \leq 2^{-n/2}$$

t follows that for each y, $\{u^n(y)\}$ is a Cauchy sequence, hence u^n onverges pointwise to a function u, which must be Borel measurable.

It remains to show that $\beta(y) = \gamma(y,u(y))$. But for any y, n,

$$\begin{aligned}\beta(y) &\in \gamma(y, U^n_{j^n(y)})\backslash\gamma(y, U^n_{j^n(y)-1}) \\ &= \gamma(y, U^n_{j^n(y)} \backslash U^n_{j^n(y)-1}) \\ &\subset \gamma(y,\overline{V^n_{j^n(y)}})\end{aligned}$$

o $\beta(y) = \gamma(y,v^n(y))$ for some $v^n(y)$ in $\overline{V^n_{j^n(y)}}$. But

$$|v^n(y) - u^n(y)| \leq \text{diam } V^n_{j^n(y)} \leq 2^{-n/2}$$

ne result follows by continuity of γ. Q.E.D.

If X is a Euclidean space then the above theorem is an easy corollary f lemma B of appendix B of Fleming and Rishel (1975). The following ore general result is found in Beneš (1970).

.5.3. Lemma. Let $(X,\underline{\underline{X}})$ be a measure space, A a separable metric pace, U a compact metric space. Let γ: X×U → A be continuous in its econd argument and measurable, $\underline{\underline{X}} \to \underline{\underline{B}}(A)$, in its first. Let β: X → A e $\underline{\underline{X}} \to \underline{\underline{B}}(A)$ measurable with $\beta(y)$ in $\gamma(y,U)$ for y in X. Then there xists a measurable function u: $(X,\underline{\underline{X}}) \to (U,\underline{\underline{B}}(U))$ such that y) = $\gamma(y,u(y))$.

6 Comments. The denseness of the control laws in the adapted ntrols, i.e. theorem 7.4, is well-known, cf. Bensoussan (1982) apter VI, Fleming and Pardoux (1982), and Krylov (1980) chapter 3, for ightly different versions.

The proof of theorem 7.4 shows that

$$\inf\{J_0(u): u \in \mathcal{U}_0\} \le \inf\{J_U(u): u \in \mathcal{U}_N\}$$

where $\mathcal{U}_N$ is the set of U-valued adapted stochastic processs defined on some $(\Omega,\underline{\underline{F}},\{\underline{\underline{F}}_t\},P,\{w_t\})$ which may vary from process to process. Since we can imbed $\mathcal{U}_0$ in $\mathcal{U}_N$, then we have equality of the infima. Note also that (7.5) could be weakened to

$$|\sigma(t,x)|^2 \le k_0(1+|x|^2).$$

8 The maximum principle

Let us consider the problem (7.14) which has no constraints. We are given $(\Omega, \underline{\underline{F}}, \{\underline{\underline{F}}_t\}, P, \{w_t\})$ and $\{u_t^*\}$ in $\tilde{\mathcal{U}}$. We are in the completely observable case so we may let $\{\underline{\underline{F}}_t\}$ be generated by the solution $\{x_t^*\}$ of (7.2). We still assume (7.3) - (7.10) [actually the condition that $\sigma(t,x)$ be onto $F(t,x)$ can be omitted] so that $|u_t^*(\omega)| \leq K^*(1+\|x^*(\omega)\|_t)$ a.s., and we also assume

(8.1) for each (t,u) f, σ, ℓ, c are continuously differentiable in x, and

$$|c_x(x)| + |\ell_x(t,x,u)| \leq k_2(1+|x|^{q-1} + |u|^q).$$

Note that the fact that we have $|x|^{q-1}$ rather than $|x|^q$ is no restriction! In fact any polynomial bound for both ℓ, ℓ_x (possibly different) suffices. Observe that (7.4) implies that f_x, σ_x are bounded by k_1.

Suppose that we are given $u^\varepsilon: [0,\varepsilon_0] \to \tilde{\mathcal{U}}$ such that $u^0 = u^*$, then

$$J_0(u^*) \leq J_0(u^\varepsilon) = J_0(u^*) + \varepsilon \frac{d}{d\varepsilon} J_0(u^\varepsilon)\Big|_{\varepsilon=0} + o(\varepsilon)$$

if the indicated derivative exists. Thus a necessary condition is that $\frac{d}{d\varepsilon} J_0(u^\varepsilon)\Big|_{\varepsilon=0} \geq 0$. We develop this idea now.

Recall the definition of $\mathcal{U}^t$, V, $\bar{q}$ of section 5. Define

$$\tilde{\mathcal{U}}^t = \{u(x^*(\cdot)): u \text{ in } \mathcal{U}^t\} \subset L^{\bar{q}}(\Omega, \underline{\underline{F}}_t, P; U).$$

For v in V, define the process $\{\tilde{v}_t\}$ by $\tilde{v}_t(\omega) = (v \circ i_t)(x^*(\omega))$ and let $\tilde{V}$ be $\{\tilde{v}: v \text{ in } V\}$. Observe that $\{\tilde{v}_t: v \text{ in } V\}$ is dense in $\tilde{\mathcal{U}}^t$, and $\tilde{V} \subset \tilde{\mathcal{U}}$. For v in $\tilde{V}$ define

$$F^v(t,x) = f(t,x,v_t) - f(t,x,u_t^*),$$

$$G^v(t,x) = \ell(t,x,v_t) - \ell(t,x,u_t^*),$$

and note that $F^v(t,x_t^*)$, $G^v(t,x_t^*)$, $|F^v(t,x_t^*)|^{\bar{q}}$ are integrable (dtdP) by (7.3), (7.5) - (7.8) and the boundedness of v. Moreover F^v is Lipschitz in x by (7.4). It follows from lemma 8.12.1 that for t not in N(v), a null set,

(8.2)
$$\frac{d}{dt}\int_0^t F^v(s,x_s^*)ds = F^v(t,x_t^*) \quad \text{a.s.}$$

$$\frac{d}{dt}\int_0^t G^v(s,x_s^*)ds = G^v(t,x_t^*) \quad \text{a.s.}$$

$$\frac{d}{dt}\int_0^t E|F^v(s,x_s^*)|^{\bar{q}}ds = E|F^v(t,x_t^*)|^{\bar{q}}.$$

We let $N_0 = \bigcup_{v \in V} N(v)$, so N_0 is also a null set.

Now fix u in $\tilde{V}$, s not in N_0. Then for each $\varepsilon > 0$ let

$$u_t^\varepsilon = \begin{cases} u_t & \text{if } s - \varepsilon < t \leq s \\ u_t^* & \text{otherwise.} \end{cases}$$

Then u^ε is in $\tilde{\mathcal{U}}$; let x^ε denote the corresponding solution of (7.2). We can now begin to compute $\frac{d}{d\varepsilon} J_0(u^\varepsilon)$.

Let $\xi_t^\varepsilon = x_t^\varepsilon - x_t^*$ and let

$$F^\varepsilon(t,x) = f(t,x,u_t^\varepsilon) - f(t,x,u_t^*),$$

$$G^\varepsilon(t,x) = \ell(t,x,u_t^\varepsilon) - \ell(t,x,u_t^*).$$

Observe that F^ε and G^ε are 0 for t not in $(s-\varepsilon,s]$ and are equal to F^u, G^u respectively for t in $(s-\varepsilon, s]$. Now $\xi_0^\varepsilon = 0$ and

(8.3)
$$d\xi_t^\varepsilon = \left[\left(f(t,x_t^*+\xi_t^\varepsilon,u_t^*) - f(t,x_t^*,u_t^*)\right) + F^\varepsilon(t,x_t^*+\xi_t^\varepsilon)\right]dt$$
$$+ \left(\sigma(t,x_t^*+\xi_t^\varepsilon) - \sigma(t,x_t^*)\right)dw.$$

We shall show that ξ^ε "almost" satisfies a linearized version of (7.2), namely (8.5) below.

8.1 <u>Lemma</u>. For all p in $[1,\bar{q}]$, $E\|\xi^\varepsilon\|_t^p = O(\varepsilon^p)$.

Proof: If the result holds for $p = \bar{q}$, then by Jensen's inequality it holds for the rest of the p. From (8.3), (7.4), the Lipschitz ontinuity of F^u and theorem 3.8.2 we obtain

$$|\xi_t^\varepsilon|^{\bar{q}} \le K\{\int_0^t |\xi_r^\varepsilon|^{\bar{q}} dr + |\int_0^t [\sigma(r,x_r^*+\xi_r^\varepsilon)-\sigma(r,x_r^*)]dw|^{\bar{q}}$$

$$+|\int_{(s-\varepsilon)\wedge t}^{s\wedge t} F^u(r,x_r^*)dr|^{\bar{q}}\},$$

$$E\|\xi^\varepsilon\|_t^{\bar{q}} \le \bar{K}\{\int_0^t E|\xi_r^\varepsilon|^{\bar{q}} dr\} + \varepsilon^{\bar{q}-1} \int_{(s-\varepsilon)\wedge t}^{s\wedge t} E|F_r^u|^{\bar{q}} dr$$

$$\le \bar{K} \int_0^t E\|\xi^\varepsilon\|_r^{\bar{q}} dr + O(\varepsilon^{\bar{q}})$$

nce by (8.2) $\frac{1}{\varepsilon}\int_{s-\varepsilon}^{s} E|F^u|^{\bar{q}} dr \to E|F^u(s,x_s^*)|^{\bar{q}} < \infty$. Gronwall's equality gives the result. Q.E.D.

te that if $\phi = \phi(t,x)$ then $\Delta\phi^\varepsilon$ denotes $\phi(t,x_t^*+\xi_t^\varepsilon) - \phi(t,x_t^*)$.

2 Corollary. If $1 \le p \le \bar{q}$ then $E\|\Delta F^\varepsilon\|_T^p = O(\varepsilon^p)$, $E\int_0^T |\Delta G^\varepsilon| dt = o(\varepsilon)$.

oof: $|\Delta F_t^\varepsilon| \le |f(t,x_t^*+\xi_t^\varepsilon,u_t^\varepsilon)-f(t,x_t^*,u_t^\varepsilon)|+|f(t,x_t^*+\xi_t^\varepsilon,u_t^*)-f(t,x_t^*,u_t^*)|$

$$\le 2k_1|\xi_t^\varepsilon|$$

the first result follows from the lemma. Now from the mean-value eorem, (8.1), (7.6) and the boundedness of u_t, it follows that

$$E\int_0^T |\Delta G^\varepsilon| dt \le K \int_{s-\varepsilon}^{s} E|\xi_t^\varepsilon|(1+\|x^*\|_t^q + |\xi_t^\varepsilon|^{q-1})dt$$

$$\le \bar{K}\,\varepsilon[E\{\|\xi^\varepsilon\|_T^q\} + E\{\|\xi^\varepsilon\|_T^{\bar{q}}\}^{1/\bar{q}}]$$

$$= o(\varepsilon).$$

nce by (7.3) (7.6) $E\|x^*\|_T^{\tilde{q}} < \infty$ for any $\tilde{q} < \infty$. Q.E.D.

Let y^ε be the solution of

.4)
$$dy^\varepsilon = [f_x(t,x_t^*,u_t^*)y^\varepsilon + F^\varepsilon(t,x_t^*)]dt + \sum_{k=1}^{d} \sigma_x^k(t,x_t^*)y^\varepsilon dw_t^k$$

$$y_0^\varepsilon = 0$$

where σ^k is the kth column of σ and w^k the kth component of w. Note that $y_t^\varepsilon = 0$ for $t \leq s-\varepsilon$.

8.3. Lemma. $E\|\xi^\varepsilon - y^\varepsilon\|_T^p = o(\varepsilon^p)$ for $2 \leq p < \bar{q}$.

Proof: Let $\tilde{y} = \xi^\varepsilon - y^\varepsilon$. By the mean-value theorem there exist processes $\{\phi_t^i\}$, $i = 0,1,\cdots,d$ with values in $[0,1]$ such that

$$d\tilde{y} = f_x(t,x_t^*,u_t^*)\tilde{y}dt + \sum_k \sigma_x^k(t,x_t^*)\tilde{y}dw_t^k + de_1 + de_2 + de_3$$

$$de_1 = [f_x(t,x_t^* + \phi_t^0\xi_t^\varepsilon,u_t^*) - f_x(t,x_t^*,u_t^*)]\xi_t^\varepsilon dt$$

$$de_2 = \sum_{k=1}^d [\sigma_x^k(t,x_t^* + \phi_t^k\xi_t^\varepsilon) - \sigma_x^k(t,x_t^*)]\xi_t^\varepsilon dw_t^k$$

$$de_3 = \Delta F^\varepsilon(t,x_t^*)dt.$$

By corollary 8.2, $E\|e_3\|_T^p = O(\varepsilon^{p+1}) = o(\varepsilon^p)$. From theorem 3.8.2 and lemma 8.1 we get

$$E\|e_2\|_T^p \leq c_0 \; E\{\|\xi^\varepsilon\|_T^p(\textstyle\int_0^T|\Delta\sigma_x|^2dt)^{p/2}\}$$

$$\leq c_1E\{\|\xi^\varepsilon\|_T^{\bar{q}}\}^{p/\bar{q}} \; E\{\textstyle\int_0^T|\Delta\sigma_x|^{p\bar{q}/(\bar{q}-p)}dt\}^{1-p/\bar{q}}$$

$$= o(\varepsilon^p)$$

since $\int_0^T E|\Delta\sigma_x|^{p\bar{q}/(\bar{q}-p)}dt \to 0$ with ε by the boundedness of σ_x and the continuity of $x \to \sigma_x(t,x)$. Similarly

$$E\|e_1\|_T^p = o(\varepsilon^p).$$

The boundedness of f_x, σ_x and Gronwall's inequality now give the result.

Q.E.D.

Let $\Phi(t,\tau)$ be the fundamental matrix solution of

$$dy = f_x(t,x_t^*,u_t^*)y\;dt + \sum_{k=1}^d \sigma_x^k(t,x_t^*)y\;dw_t^k, \tag{8.5}$$

i.e. $\Phi^\ell(\cdot,\tau)$ is a solution of (8.5) on $t > \tau$ and $\Phi(\tau,\tau) = I$, the $n\times n$ identity matrix.

8.4 Lemma. For $p < \infty$, $E|\Phi(t,\tau)|^p$ is bounded uniformly in t,τ.

Proof: This follows from (8.5) and the fact that f_x and σ_x are bounded. Q.E.D.

8.5 Lemma. If $T \geq t \geq s$, then for any p in $[2,\bar{q})$

$$E|y_t^\varepsilon - \varepsilon\Phi(t,s)F^u(s,x_s^*)|^p = o(\varepsilon^p)$$

uniformly in t.

Proof: Since $F^\varepsilon(t,x_t^*) = 0$ for $t > s$ then $y_t^\varepsilon = \Phi(t,s)y_s^\varepsilon$. Moreover

$$y_s^\varepsilon = \int_{s-\varepsilon}^s f_x(r,x_r^*,u_r^*)y_r^\varepsilon dr + \sum_k \int_{s-\varepsilon}^s \sigma_x^k(r,x_r^*)y_r^\varepsilon dw_r^k + \int_{s-\varepsilon}^s F^u(r,x_r^*)dr. \tag{8.6}$$

Let $2 \leq p < \bar{q}$. From theorem 3.8.2 and lemmas 8.1, 8.3 it follows that

$$E|\int_{s-\varepsilon}^s f_x y^\varepsilon dr + \sum_k \int_{s-\varepsilon}^s \sigma_x^k y^\varepsilon dw^k|^p = O(\varepsilon^{p/2})E\|y^\varepsilon\|_s^p = o(\varepsilon^p). \tag{8.7}$$

Also

$$E|\int_{s-\varepsilon}^s F^u(r,x_r^*)dr - \varepsilon F^u(s,x_s^*)|^p = o(\varepsilon^p) \tag{8.8}$$

by (8.2), the Lebesgue convergences theorem and the fact that

$$E|\frac{1}{\varepsilon}\int_{s-\varepsilon}^s F^u dr|^{\bar{q}} \leq \frac{1}{\varepsilon}\int_{s-\varepsilon}^s E|F^u|^{\bar{q}}dr$$

i.e. the left side is bounded uniformly in ε. Now (8.6),(8.7),(8.8) and lemma 8.4 imply

$$E|y_t^\varepsilon - \varepsilon\Phi(t,s)F^u(s,x_s^*)|^p \leq E|\Phi(t,s)|^p|y_s^\varepsilon - \varepsilon F^u(s,x_s^*)|^p = o(\varepsilon^p).$$

Q.E.D.

From lemmas 8.3 and 8.5 we obtain the promised conclusion.

8.6 Corollary.

$$\xi_t^\varepsilon = \varepsilon\,\Phi(t,s)F^u(s,x_s^*) + \bar{\rho}_t \text{ if } t \geq s,$$

where $$\sup_t E|\bar{\rho}_t|^p = o(\varepsilon^p).$$

We return now to the functional J_0 and we observe that

$$J_0(u^\varepsilon) - J_0(u^*) = E\{\int_0^T[\ell(t,x_t^*+\xi_t^\varepsilon,u_t^*) - \ell(t,x_t^*,u_t^*) + G^\varepsilon(t,x_t^*+\xi_t^\varepsilon)]dt$$
$$+ [c(x_T^*+\xi_T^\varepsilon) - c(x_T^*)]\}.$$

8.7 Lemma. $J_0(u^\varepsilon)-J_0(u^*)=E\{\int_0^T[\ell_x(t,x_t^*,u_t^*)\xi_t^\varepsilon+G^\varepsilon(t,x_t^*)]dt+c_x(x_T^*)\xi_T^\varepsilon\}+E\rho_1$ where $E|\rho_1| = o(\varepsilon)$.

Proof: By the mean value theorem, for some ϕ_t in $[0,1]$,

$$\rho_1 = \int_0^T\{[\ell_x(t,x_t^*+\phi_t\xi_t^\varepsilon,u_t^*) - \ell_x(t,x_t^*,u_t^*)]\xi_t^\varepsilon + \Delta G^\varepsilon\}dt$$
$$+ [c_x(x_T^*+\phi_T\xi_T^\varepsilon) - c_x(x_T^*)]\xi_T^\varepsilon.$$

Now let p be such that $\bar{q}(q-1)^{-1} > p > \bar{q}(\bar{q}-1)^{-1}$; then $p(p-1)^{-1} < \bar{q}$, and

$$E|\rho_1| \le K\ E\{\|\xi^\varepsilon\|_T^{p/p-1}\}^{\frac{p-1}{p}}\ E\{\int_0^T|\Delta\ell_x^\varepsilon|^p dt + |\Delta c_x^\varepsilon|^p\}^{1/p} + o(\varepsilon)$$

using corollary 8.2.

By lemma 8.1, the first factor on the right side is $O(\varepsilon)$. We show now that the second factor is $o(1)$. If $\varepsilon_n \to 0$ then by lemma 8.1 for any subsequence ε_{n_k} there exists a further subsequence, again called ε_{n_k}, such that $\|\xi^{\varepsilon_{n_k}}\|_T \to 0$ a.s., and hence $|\Delta\ell_x^{\varepsilon_{n_k}}| \to 0$ a.e. (dtdP) since ℓ_x is continuous in x. Moreover $\bar{q}(q-1)^{-1} > p$, so by (7.6) and (8.1)

$$E\int_0^T|\Delta\ell_x^\varepsilon|^{\bar{q}/(q-1)}dt \le K\ E\{1+\|x^*\|_T^{q\bar{q}/(q-1)} + \|\xi^\varepsilon\|_T^{\bar{q}}\} < \bar{K}$$

for some constant $\bar{K}$ for all ε such that $E\|\xi^\varepsilon\|_T^{\bar{q}} < 1$ (i.e. ε small enough, cf. lemma 8.1). Thus by uniform integrability $E\int_0^T|\Delta\ell_x^{\varepsilon_{n_k}}|^p dt \to 0$, and so $E\int_0^T|\Delta\ell_x^\varepsilon|^p dt \to 0$. Similarly for $|\Delta c_x^\varepsilon|^p$.

Q.E.D.

We remark that (8.2) implies

$$\int_0^T G^\varepsilon(t,x_t^*)dt = \int_{s-\varepsilon}^s G^u(t,x_t^*)dt = \varepsilon G^u(s,x_s^*) + \rho_2 \tag{8.9}$$

with $E|\rho_2| = o(\varepsilon)$.

8.8 Lemma. $J_0(u^\varepsilon)-J_0(u^*) = \varepsilon E\{[\int_s^T\ell_x(t,x_t^*,u_t^*)\Phi(t,s)dt$

$$+ c_x(x_T^*)\Phi(T,s)]F^u(s,x_s^*) + G^u(s,x_s^*)\} + o(\varepsilon).$$

Proof: From lemma 8.7, and corollary 8.6, and from (8.9) it follows that we need only show that $E \int_0^s \ell_x(t,x_t^*,u_t^*)y_t^\varepsilon dt = o(\varepsilon)$. But

$$E \int_0^s |\ell_x(t,x_t^*,u_t^*)y_t^\varepsilon| dt$$

$$= E \int_{s-\varepsilon}^s |\ell_x(t,x_t^*,u_t^*)| \, |y_t^\varepsilon| dt$$

$$\leq \{E \int_{s-\varepsilon}^s |\ell_x|^2 dt \; E\|y^\varepsilon\|_T^2 \; \varepsilon\}^{1/2}$$

$$= o(\varepsilon)$$

since $\int E|\ell_x|^2 dt < \infty$, $E\|y^\varepsilon\|_T^2 \leq 2\,E\|\xi^\varepsilon\|_T^2 + 2\,E\|\xi^\varepsilon - y^\varepsilon\|_T^2 = O(\varepsilon^2)$ by lemmas 8.1 and 8.3.

Q.E.D.

8.9 Theorem. Assume (7.3), (7.4), (7.6) - (7.10) and (8.1). If u^* solves (7.14) then there exits a Lebesgue null set N_0 such that for t not in N_0,

$$\max_{u \in U} H(t,x_t^*,u,p_t) = H(t,x_t^*,u_t^*,p_t) \quad \text{a.s.}$$

where

$$p_t' = -E\{\int_0^T \ell_x(s,x_s^*,u_s^*)\Phi(s,t)ds + c_x(x_T^*)\Phi(T,t) | \underline{\underline{F}}_t\}$$

$$H(t,x,u,p) = p'f(t,x,u) - \ell(t,x,u).$$

Proof: Let $\bar{p}_t' = -\{\int_0^T \ell_x(s,x_s^*,u_s^*)\Phi(s,t)ds + c_x(x_T^*)\Phi(T,t)\}$. Then lemma 8.8 says that

$$J_0(u^\varepsilon) - J_0(u^*) = \varepsilon\, E\{H(s,x_s^*,u_s,\bar{p}_s) - H(s,x_s^*,u_s^*,\bar{p}_s)\} + o(\varepsilon)$$

hence for t not in N_0 and u in $\tilde{V}$

(8.10) $$E\, H(t,x_t^*,u_t,\bar{p}_t) \leq E\, H(t,x_t^*,u_t^*,\bar{p}_t).$$

We will first extend (8.10) to all u_t in $\tilde{\mathcal{U}}^t$.

Given v in $\tilde{\mathcal{U}}^t$ it follows that there exist u^n in $\tilde{V}$ such that $u_t^n \to v$ in $L^{\bar{q}}(\Omega,\underline{\underline{F}}_t,P;U)$, hence (by possibly taking a subsequence) $u_t^n \to v$ a.s. Since $u_t^n \to v$ in $L^{\bar{q}}$, hence is bounded in $L^{\bar{q}}$, then $E|\ell(t,x_t^*,u_t^n)|^{\bar{q}/q}$ is bounded uniformly in n, and

$$E(|\bar{p}'f(t,x_t^*,u_t^n|)^{\bar{q}/2} \le \{E|\bar{p}_t|^{\bar{q}} \; E|f(t,x_t^*,u_t^n)|^{\bar{q}}\}^{1/2}$$

is also bounded since $E|\bar{p}_t|^{\bar{q}} < \infty$. By uniform integrability $E|H(t,x_t^*,u_t^n,\bar{p}_t) - H(t,x_t^*,v,\bar{p}_t)| \to 0$. Hence (8.10) holds for all u_t in $\tilde{U}^t$.

As in corollary 6.2 this implies that for u in U

$$E\{H(t,x_t^*,u,\bar{p}_t)|\underline{\underline{F}}_t\} \le E\{H(t,x_t^*,u_t^*,\bar{p}_t)|\underline{\underline{F}}_t\} \quad \text{a.s.}$$

i.e.

$$H(t,x_t^*,u,E\{\bar{p}_t|\underline{\underline{F}}_t\}) \le H(t,x_t^*,u_t^*,E\{\bar{p}_t|\underline{\underline{F}}_t\}) \quad \text{a.s.} \qquad \text{Q.E.D.}$$

If we combine theorem 8.9 with corollary 6.7 then we obtain a complete maximum principle. First we gather the assumptions together.

(A) $f(t,x,u)$, $\ell(t,x,u)$ are continuous in u for each (t,x), and $f(t,x,u)$, $\sigma(t,x)$, $\ell(t,x,u)$, $c(x)$ are continuously differentiable as functions of x for each (t,u), with

$$|f_x| + |\sigma_x| \le k_1,$$

$$|f|^2 + |\sigma|^2 \le k_1(1+|x|^2+|u|^2),$$

$$\sigma(t,x): \mathbf{R}^d \to \mathbf{R}^n \text{ is onto } F(t,x) \text{ for each } (t,x),$$

$$|\sigma(t,x)^+| \le k_0, \; \int_0^T|\sigma(t,x)|^2 dt \le k_0,$$

$$|\ell| + |c| \le k_2(1+|x|^q+|u|^q) \text{ for some } q \text{ in } [1,\infty),$$

$$|\ell_x| + |c_x| \le k_2(1+|x|^{q-1}+|u|^q),$$

$E\exp\{\varepsilon|x_0|^2\} < \infty$ for some $\varepsilon > 0$, U is closed, and $\mathcal{U}$ contains the set of measurable, $\{\underline{\underline{G}}_t^n\}$-adapted functions $u:[0,T]\times C^n \to U$ such th

$$|u(t,x)| \le K_u(1+\|x\|_t).$$

8.10 Theorem. Assume (A). If u^* in $\mathcal{U}$ satisfies $|u^*(t,x)| \le K^*(1+\|x\|_t)$, and u^* solves (3.4), then there exists λ in $\mathbf{R}^{m_1+1+m_2}$, $\lambda \ne 0$, $\lambda_i \le 0$ if $i \ge 0$, $\lambda_i J_i(u^*) = 0$ if $i > 0$, and a null set N_0, such that for t in $(0,T]\backslash N_0$

(8.11) $$\max_{u\in U} H(t,x_t,u,p_t,\lambda) \le H(t,x_t,u_t^*,p_t,\lambda) \quad \text{a.s.}$$

where $H(t,x,u,p,\lambda)$ is given by (6.1) and

(8.12) $$p'_t = \lambda' \, E\{c_x(x_T)\Phi(T,t) + \int_t^T \ell_x(s,x_s,u_s^*)\Phi(s,t)ds \,|\, \underline{\underline{F}}_t^x\}.$$

Note that we have gone back to the convention of writing x_t for the state process corresponding to u^*.

8.11 Remarks.

8.11.1 If $u^*(t,x) = v^*(t,x_t)$, i.e. u^* is Markov, then x is a Markov process and by the Markov property we can replace $\underline{\underline{F}}_t^x$ by the σ-algebra generated by x_t in the above definition of p_t. But observe that the hypothesis (A) does not allow us to consider u^* which is optimal only among the Markov controls; see however Haussmann (1981) for that case.

8.11.2 We have not found a representation for $\{\chi_t\}$ so that we cannot apply theorem 6.4 to obtain sufficiency of the above necessary conditions.

8.11.3 The result is homogeneous in λ so if $\lambda_0 \neq 0$, then we can rescale to set $\lambda_0 = -1$.

8.12 Appendix

8.12.1 Lemma. Let $\phi:[0,T]\times\Omega \to \mathbf{R}$ be measurable. If $\int_0^T|\phi(t,\omega)|dt < \infty$ then there exists a null set N such that the function $\Phi(t,\omega) = \int_0^t\phi(s,\omega)ds$ is differentiable in t and for t not in N

$$\frac{d}{dt}\Phi(t,\omega) = \phi(t,\omega) \quad \text{a.s.}$$

Proof: Let $F(t,\omega,h) = [\Phi(t+h,\omega) - \Phi(t,\omega)]/h$, $h \neq 0$. Since F is continuous in h then

$$\liminf_{h\to 0} F = \liminf_{\substack{h\to 0\\ h \text{ rational}}} F$$

and similarly for lim sup. Hence these functions, being countable limits of measurable functions, are measurable, and thus

$$S = \{(t,\omega): \liminf_{h\to 0} F(h,t,\omega) = \limsup_{h\to 0} F(h,t,\omega)\}$$

is measurable.

Since $\Phi = \int_0^t \phi\, ds$ then a.s. $\frac{d\Phi}{dt} = \phi(t,\omega)$ for t not in N_ω, a Lebesgue null set. But S_ω, the ω-section of S, must contain $[0,T]\setminus N_\omega$, so must have full measure. Fubini's theorem then implies that S has full measure, and hence for almost all t, S_t has full measure, that is $P(S_t) = 1$.

Q.E.D.

8.13 Exercises. The first four exercises develop a maximum principle for the case when the controls are adapted processes and σ may be degenerate. The basic assumptions are:

$(\Omega, \underline{\underline{F}}, \{\underline{\underline{F}}_t\}_{0\leq t\leq T}, P)$ is a filtered probability space, $\underline{\underline{F}}$ is countably generated, U is a Borel set in $\mathbf{R}^m$, $\{w_t : 0\leq t\leq T\}$ is a standard Brownian motion,

$$|f(t,x,u)|^2 + |\sigma(t,x)|^2 \leq k_1(1+|x|^2)$$

$$|f_x(t,x,u)| + |\sigma_x(t,x)| \leq k_1$$

$$|c(x)| + |c_x(x)| + |\ell(t,x,u)| + |\ell_x(t,x,u)| \leq k_2(1+|x|^q+|u|^q),$$

the indicated derivatives existing and being continuous (in x). Also f, ℓ are continuous in u for each (t,x).

$$E|x_0|^p < \infty \quad \text{for any } 1 \leq p < \infty.$$

Take $\bar{q} > \max\{2,q\}$. N.b. $\underline{\underline{F}}_{t-}$ is the smallest σ-algebra containing $\underline{\underline{F}}_s$, $s < t$. Let $\mathcal{U}$ be the set of all $\{\underline{\underline{F}}_{t-}\}$ adapted stochastic processes such that $\sup_t E|u_t|^{\bar{q}} < \infty$. For u in $\mathcal{U}$, let $\{x_t^u\}$ denote the unique strong solution of

$$dx = f(t,x,u_t)dt + \sigma(t,x)dw_t.$$

8.13.1 Let V be a countable dense subset of $L^{\bar{q}}(\Omega, \underline{\underline{F}}, P; U)$ and let

$$V^t = \{E\{u|\underline{\underline{F}}_{t-}\} : u \text{ in } V\}.$$

Show that V^t is dense in $L^{\bar{q}}(\Omega, \underline{\underline{F}}_{t-}, P; U)$, and that for v in V^t, $E\{v|\underline{\underline{F}}_s\} \to v$ a.s. as $s \uparrow t$.

Note: Doob (1953) chapter VII, §11, shows that

$$\lim_{s\uparrow t} E\{v|\underline{\underline{F}}_s\} = E\{v|\underline{\underline{F}}_{t-}\}.$$

8.13.2 Assume u^* is in $\mathcal{U}$ and x^* is x^{u^*}. Let D be the convex cone in $\mathbf{R}^{m_1+1+m_2}$ generated by

$\{Ec_x(x_T^*)\Phi(T,s)[f(s,x_s^*,u) - f(s,x_s^*,u_s^*)]: 0 < s \leq T, \text{ u in } V^s\}$.

Show that D is a cone of variations of J at u^* if $J(u) = E\{c(x_T^u)\}$.

Hint: in the notation of theorem 5.4, define

$$u_t^\rho = \begin{cases} E\{v_i | \underline{\underline{F}}_{t_i-\tau_i}\} & \text{if t in } I_i(\rho),\ i = 1,\cdots,r \\ u_t^* & \text{otherwise.} \end{cases}$$

13.3 Assume that u^* solves

$$\inf\{J_0(u): \text{u in } \mathcal{U},\ J_i(u) = 0 \text{ if } i < 0,\ J_i(u) = 0 \text{ if } i > 0\}$$

with $J(u) = E\{c(x_T^u)\}$. Show that there exists λ in $\mathbf{R}^{m_1+1+m_2}$, $\lambda \neq 0$, $\lambda_i \leq 0$ if $i \geq 0$, $\lambda_i J_i(u^*) = 0$ if $i \neq 0$, such that for almost all t, for all u in U

.13)
$$\lambda' E\{c_x(x_T^*)\Phi(T,t) | \underline{\underline{F}}_{t-}\} f(t,x_t^*,u) \leq \lambda' E\{c_x(x_T^*)\Phi(T,t) | \underline{\underline{F}}_{t-}\} f(t,x_t^*,u_t^*) \text{ a.s.}$$

13.4 Show that if in 8.13.3,

$$J(u) = E\{\int_0^T \ell(t,x_t^u,u_t)dt + c(x_T^u)\}$$

then (8.13) becomes

$$\lambda' E\{c_x(x_T^*)\Phi(T,t) - \int_t^T \ell_x(s,x_s^*,u_s^*)\Phi(s,t)ds | \underline{\underline{F}}_{t-}\} f(t,x_t^*,u) + \lambda'\ell(t,x_t^*,u)$$
$$\leq \lambda' E\{c_x(x_t^*)\Phi(T,t) + \int_t^T \ell_x(s,x_s^*,u_s^*)\Phi(s,t)ds | \underline{\underline{F}}_{t-}\} f(t,x_t^*,u_t^*) + \lambda'\ell(t,x_t^*,u_t^*)$$

a.s.

13.5 Show that if the assumptions A hold and if u^* solves

$$\inf\{J_0(u): \text{u in } \mathcal{U},\ E\, c_1(x_{t_1}) \leq 0\}$$

with $0 < t_1 < T$, then the maximum principle holds where the adjoint process is given by (8.12) with $c_x(x_T)\Phi(T,t)$ replaced by

$$\begin{bmatrix} \nabla c_0(x_T)\ \Phi(T,t) \\ 0 \end{bmatrix} + \begin{bmatrix} 0 \\ \nabla c_1(x_{t_1}) \end{bmatrix} \Phi(t_1,t) 1_{t_1}(t).$$

N.b. Here $m_1 = 0$, $m_2 = 1$, $1_{t_1}(t)$ is the characteristic function of $[0,t_1)$. Also $\ell = \begin{bmatrix} \ell_0 \\ 0 \\ 0 \end{bmatrix}$. You must find the correct cone of variations, etc.

8.13.6 Show that if the assumptions A hold and u^* solves

$$\inf\{J_0(u): u \text{ in } \mathcal{U},\ E\, c_1(x_{t_1}, E\, x_{t_1}) \leq 0\}$$

where $c_1 = c_1(x,y)$ with

$$|c_1| + |(c_1)_x| + |(c_1)_y| \leq k_2(1+|x|^q+|y|^q)$$

then the maximum principle holds where the adjoint process is given as in problem 8.13.5 except $\nabla c_1(x_{t_1})$ is replaced by

$\nabla_x c_1(x_{t_1}, E\, x_{t_1}) + \nabla_y c_1(x_{t_1}, E\, x_{t_1})$.

<u>Hint</u>: $c(x_{t_1}, E^u x_{t_1}) = c(x_{t_1}, Ex_{t_1}) + c_y(x_{t_1}, E_{x_{t_1}})(E^u x_{t_1} - Ex_{t_1})$

$+\, o(|E^u x_{t_1} - Ex_{t_1}|)$, and we can apply theorem 3.8.3 to

$E\{c(x_{t_1}, Ex_{t_1}) | \underline{F}_t\}$ and to $E\{x_{t_1} | \underline{F}_t\}$.

8.14 <u>Comments</u>. The linearization technique employed in this section was used by Kushner (1972) to derive a maximum principle for controls which are adapted processes (rather than feedback controls). More general constraints can be employed; for example, with $0 < t_i < T$,

$$E\, c_i(x_{t_i}) \leq (=) 0$$

will give rise to a discontinuity in the adjoint process. For details see Kushner (1972) and Haussmann (1979). The problem when $c(x)$ is replaced by $c(x, Ex)$ was treated by Kushner (1972) for controls which are adapted processes, and by Ichikawa (1978) for feedback controls. Lemma 8.12.1 can be found in Kushner (1972).

9 Examples

After having worked so hard to obtain our result (theorem 8.10), it would be pleasant if we were able to apply it to solve a stochastic control problem, in other words, to find an optimal control. We show now that this is indeed the case - with a small caveat. Since the conditions are only necessary, but not necessarily sufficient, we can, in fact, only exhibit controls which satisfy these necessary conditions. Beyond this section we shall endeavour to show that these controls are optimal. Let us begin with a definition.

9.1 Definition. The control u^* in $\mathcal{U}$ is extremal for the problem (3.4) if it is admissible, and if

i) $|u^*(t,x)| \leq K^*(1+\|x\|_t)$ for some $K^* < \infty$

ii) there exists λ in $\mathbf{R}^{m_1+1+m_2}$, $\lambda \neq 0$, $\lambda_i \leq 0$ if $i \geq 0$,

$\lambda_i J_i(u^*) = 0$ if $i > 0$

iii) there exists a null set N_0 such that for

t in $(0,T]\backslash N_0$

(9.1) $$\max_{u\in U} H(t,x_t,u,p_t,\lambda) = H(t,x_t,u_t^*,p_t,\lambda) \quad \text{a.s.}$$

where

(9.2) $$H(t,x,u,p,\lambda) = \lambda'\ell(t,x,u) + p'f(t,x,u),$$

(9.3) $$p_t' = \lambda' E\{c_x(x_T)\Phi(T,t) + \int_t^T \ell_x(s,x_s,u_s^*)\Phi(s,t)ds \,|\, \underline{\underline{F}}_t^x\},$$

(9.4) $$d\,\Phi(s,t) = f_x(s,x_s,u_s^*)\Phi(s,t)ds + \sum_k \sigma_x^k(s,x_s)\Phi(s,t)dw_s^k,\ \Phi(t,t) = I.$$

The reader should compare the definition of "u^* extremal" with "u^* strongly extremal", definition 6.3. We can now restate theorem 8.10 as:

Assume (A). If u^* is optimal then it is extremal.

Finally we remark that although condition (A) fails for Markovian controls since (A) requires $\mathcal{U}_G \subset \mathcal{U}$ which is false if $\mathcal{U}$ is the set of Markovian controls; nevertheless if an optimal u^* is Markovian then it is certainly optimal within the smaller class of Markovian controls.

9.2 The linear regulator. Here we set

$\mathcal{U} = \{u:[0,T]\times C^n \to \mathbf{R}^m$, measurable, $\{\underline{\underline{G}}^n_t\}$ adapted such that (9.5) has a weak solution$\}$.

(9.5)
$$dx_t = [A(t)x_t + B(t)u_t]dt + C(t)dw_t, \quad x_0 \text{ fixed},$$
$$J(u) = E\{\int_0^T [x_t'M(t)x_t + u_t'N(t)u_t]dt + x_T'Dx_T\}$$

We assume that $U = \mathbf{R}^m$, A,B,C,M,N are bounded, measurable, $M(t) \geq 0$, $D(t) \geq 0$, $N(t) > 0$, $N(t)^{-1}$ bounded, $C(t)^{-1}$ exists and is bounded on the range of B(t) and M(t), N(t), D(t) are symmetric. Now note that if u satisfies

$$u(t,x) = K_u(1+\|x\|_t)$$

then a weak solution of (9.5) can be obtained by a Girsanov transformation from the strong solution of (9.5) with u = 0 (which always exists). Hence $\mathcal{U}_G \subset \mathcal{U}$ and (A) holds.

We proceed to look for an extremal control. The first step is to write the analogues of (9.2)-(9.4).

(9.2)' $$H(t,x,u,p,\lambda) = p'(Ax + Bu) + \lambda'(x'Mx+u'Nu),$$

(9.3)' $$p_t' = 2\lambda\{\bar{x}_{Tt}'D\,\Phi(T,t) + \int_t^T \bar{x}_{st}'M(s)\Phi(s,t)ds\}$$

(9.4)' $$\frac{d}{ds}\Phi(s,t) = A(s)\Phi(s,t), \; \Phi(t,t) = I$$

where $\bar{x}_{st} = E\{x_s|\underline{\underline{F}}^x_t\}$. Since $\lambda = \lambda_0 \neq 0$ we may take $\lambda = -1$, and so H is concave in u. The inequality (9.1) is therefore equivalent to

(9.1)' $$u_t^* = N(t)^{-1}B(t)'p_t/2,$$

and a control u^* is extremal if we can find it in the form $u^*(t,x)$ to satisfy (9.1)' with p_t given by (9.3)'.

At this point some intuition is required which we hope to motivate by the following argument. From (9.3)' we see that p_t is linear in $\bar{x}_{\cdot t}$, which in turn satisfies

(9.6) $$\bar{x}_{st} = \Phi(s,t)x_t + \int_t^s \Phi(s,\tau)B(\tau)\, E\{u_\tau^*|\underline{\underline{F}}^x_t\}d\tau.$$

Thus it follows from (9.1)' that u_t^* is linear in $\{\bar{x}_{st}: s \geq t\}$, i.e. linear in $(x_t, E\{u_\cdot^*|\underline{\underline{F}}^x_t\})$, and whence it seems reasonable to take $E\{u_\cdot^*|\underline{\underline{F}}^x_t\}$ linear in x_t, and thus u_t^* linear in x_t. Hence let us attempt to show that

$$u^*(t,x) = K(t)x_t$$

is extremal for suitable K(t).

We first observe that in this case the Markov property entails $E\{\cdot|\underline{\underline{F}}_t^x\} = E\{\cdot|x_t\}$, and (9.5) has a strong solution if K(t) is bounded. Hence for bounded K(t), u^* is admissible. Let $\Phi_K(s,t)$ denote the fundamental matrix solution of

$$\frac{dx}{dt} = [A(t) + B(t)K(t)]x.$$

Then in (9.3)'

$$\bar{x}_{st} = \Phi_K(s,t)x_t \tag{9.7}$$

$$p_t = -2P(t)x_t$$

if we define

$$P(t) = \Phi'(T,t)D\Phi_K(T,t) + \int_t^T \Phi'(s,t)M(s)\Phi_K(s,t)ds. \tag{9.8}$$

But now (9.1)' is equivalent to

$$u^*(t,x) = -N(t)^{-1}B(t)'P(t)\ x_t \tag{9.9}$$

which means

$$K(t) = -\ N(t)^{-1}B(t)'P(t)$$

with P(t) given by (9.8), or, if we differentiate (9.8), by the Riccati equation

$$\frac{dP}{dt} + A(t)'P(t) + P(t)A(t) - P(t)B(t)N(t)^{-1}B(t)'P(t) + M(t) = 0 \tag{9.10}$$

$$P(T) = D.$$

We can conclude that P, hence K, is bounded. Indeed a bit of manipulation shows

$$\frac{dP}{dt} + [A(t)+B(t)K(t)]'P(t) + P(t)[A(t)+B(t)K(t)] + K(t)'N(t)K(t) + M(t) = 0$$

so that

$$P(t) = \Phi_K(T,t)'D\Phi_K(T,t) + \int_t^T \Phi_K(s,t)'[M(s)+K(s)'N(s)K(s)]\Phi_K(s,t)ds$$

$$= \Phi(T,t)'D\Phi(T,t) + \int_t^T \Phi(s,t)'[M(s)-P(s)'B(s)N(s)^{-1}B(s)'P(s)]\Phi(s,t)ds$$

The first of these two equalities implies that $P(t) \geq 0$, while the second shows that

$$P(t) \leq \Phi(T,t)'D\Phi(T,t) + \int_t^T \Phi(s,t)'M(s)\Phi(s,t)ds,$$

hence is bounded. Thus u^* given by (9.9) is extremal! We shall see later that it is optimal.

Let us now add a constraint:

$$E|x_T|^2 - \beta \leq 0.$$

In this case

(9.2)" $\quad H(t,x,u,p,\lambda) = p'[A(t)x + B(t)u] + \lambda_0[x'M(t)x + u'N(t)u]$

(9.3)" $\quad p_t' = 2\lambda_0\{\bar{x}_{Tt}' D\Phi(T,t) + \int_t^T \bar{x}_{st}' M(s)\Phi(s,t)ds\} + 2\lambda_1\bar{x}_{Tt}'\Phi(T,t).$

We must consider two cases. In the normal case, $\lambda_0 \neq 0$, we can renormalize λ to obtain $\lambda = (-1,-\rho)$ with $\rho \geq 0$ and so (9.3)" is

$$p_t' = -2\{\bar{x}_{Tt}' D_\rho\Phi(T,t) + \int_t^T \bar{x}_{st}' M(s)\Phi(s,t)ds\}$$

with $D_\rho = D + \rho I \geq 0$. Moreover (9.1) is still equivalent to (9.1)', so that an extremal must also be an extremal for the unconstrained problem with D replaced by D_ρ. We know that an extremal is

$$u^\rho(t,x) = - N(t)^{-1}B(t)'P_\rho(t)x_t$$

where we have denoted the dependence of P on ρ (due to D_ρ) by the subscript ρ. Now ρ must be chosen so as to satisfy the remaining necessary condition, i.e. $\lambda_1 J_1(u^*) = 0$, as well as the constraint $J_1(u^*) \leq 0$. It may happen that the solution of the original problem (9.5) without constraints (i.e. $\rho = 0$) already satisfies

$$E|x_T|^2 \leq \beta$$

in which case we simply take $\rho = 0$, i.e. $\lambda = (-1,0)$. If however this is not the case we must choose ρ such that $E|x_T^\rho|^2 = \beta$ (x^ρ is the solution corresponding to u^ρ) since then $\rho \neq 0$ and we require $-\rho J_1(u^*) = 0$. We must ascertain whether this can be done for $\rho \geq 0$. Set

$$\begin{aligned} R(\rho) &= E|x_T^\rho|^2 \\ &= x_0'\Phi_\rho(T,0)\Phi_\rho(T,0)x_0 + \text{trace}\int_0^T \Phi_\rho(T,s)'C(s)C(s)'\Phi_\rho(T,s)ds \end{aligned}$$

where $\Phi_\rho = \Phi_{K_\rho}$. If $R(0) \leq \beta$ then we know that the extremal control for the unconstrained problem is extremal here also. Note that $D_\rho = D + \rho I$ represents D plus a penalization factor $\rho|x_T|^2$. It is clear that as $\rho \uparrow$

then $E|x_T^\rho|^2 \downarrow$, that is $R(\rho) \downarrow$ so that R_∞ must exist. If $R(0) > \beta > R(\infty) = \lim_{\rho\to\infty} R(\rho)$ then there exists ρ^* such that $R(\rho^*) = \beta$ and $u^* = u^{\rho^*}$ is extremal with $\lambda = (-1,-\rho^*)$. Finally if $\beta \leq R_\infty$ then no extremal control exists - in fact no admissible control exists (with the property that $u_t \leq K_u(1+\|x\|_t)$). That such a possibility may exist is seen from the degenerate example $B(t) = 0$. Here

$$R(\rho) = E|x_T|^2$$

is independent of ρ, i.e. is constant, and may exceed β if $|x_0|$ is sufficiently large.

Let us now turn to the abnormal case $\lambda_0 = 0$. We renormalize $\lambda = (0,-1)$. Here

$$H(t,x,u,p,\lambda) = p'\{A(t)x+B(t)u]$$

which attains a maximum on $\mathbf{R}^m$ only if $p'B(t) = 0$, in which case u is arbitrary. Hence no abnormal extremals exist unless, cf. (9.3)",

$$\bar{x}'_{Tt}\Phi(T,t)B(t) = 0 \quad \text{a.e.}(t) \tag{9.11}$$

and u^* can be chosen such that $E|x_T|^2 = \beta$ (since $\lambda_1 J_1(u^*) = 0$). We can show that if (A,B) are controllable then no abnormal extremals exist. Indeed assume that for all $s < t$

$$W(t,s) = \int_s^t \Phi(t,r)B(r)B(r)'\Phi(t,r)'dr > 0$$

i.e. for any x_0, y there exists u in $L^2(s,t)$ such that, if x is the solution of

$$dx = (Ax + Bu)dt, \quad x(s) = x_0,$$

then $x(t) = y$. Then (9.11) implies

$$\bar{x}'_{Ts}\Phi(T,t)B(t) = 0, \quad s \leq t$$

$$\text{i.e.} \quad \bar{x}'_{Ts}\, W(T,s)\bar{x}_{Ts} = 0$$

$$\text{i.e.} \quad \bar{x}_{Ts} = E\{x_T|\underline{F}_s^x\} = 0.$$

But this is impossible (unless $x_0 = 0$, and the range of A(t) is contained in the range of B(t), a.e.(t)).

If we drop the controllability hypothesis, abnormal extremals are possible, as the case $B(t) = 0$ shows (but only for very special x_0).

9.3 The Beneš-Hilborn predicted miss problem. For this example we set

$$\mathcal{U} = \{u: [0,T]\times C^n \to U, \text{ measurable } \{\underline{\underline{G}}^n_t\} \text{ adapted such that (9.5) has a weak solution}\}$$

$$U = \{u \text{ in } \mathbf{R}^m: |u_i| \leq 1\},$$

$$J(u) = E\, k(v\cdot x_T)$$

where the assumptions on A,B,C made in example 9.2 are still assumed to hold, and $k:\mathbf{R} \to \mathbf{R}$ is even, non-negative, continuously differentiable, monotone increasing for $y > 0$, and

$$|k(y)| + |k_y(y)| \leq k_2(1+|y|^q)$$

where $k_y = dk/dy$. Since U is bounded then $\mathcal{U}_G \subset \mathcal{U}$ and (A) holds. Now

$$H(t,x,u,p,\lambda) = p'\big(A(t)x + B(t)u\big)$$

$$p'_t = E\{\lambda' k_y(v\cdot x_T)v'\Phi(T,t)|\underline{\underline{F}}^x_t\}$$

$$d\Phi(s,t) = A(s)\Phi(s,t)ds,\ \Phi(t,t) = I.$$

As there are no constraints we may take $\lambda = \lambda_0 = -1$. The Hamiltonian inequality (9.1) is equivalent to

$$u^*_t = \operatorname{sgn}[B(t)'p_t]$$

where $\operatorname{sgn} a = a|a|^{-1}$ if a is in $\mathbf{R}$, and if a is in $\mathbf{R}^m$ then $(\operatorname{sgn} a)_i = \operatorname{sgn}(a_i)$. We set sgn 0 = 0 arbitrarily. Observe that if we set

$$s(t) = \Phi(T,t)'v,\quad y_t = x_t\cdot s(t)$$

then u^* is extremal if

(9.12) $$u^*_t = -\operatorname{sgn} E\{k_y(y_T)|\underline{\underline{F}}^x_t\}\ \operatorname{sgn}[B(t)'s(t)].$$

So the task of finding an extremal u^* reduces to finding $u^* = u^*(t,x)$ in $\mathcal{U}$ such that (9.12) holds.

Again a small amount of intuition is required to produce an educated guess. Since k has a global minimum at zero, then we really wish to minimize $|v\cdot x_T| = |y_T|$. But

$$ds = -A(t)'s\,dt$$

so

$$dy_t = s(t)'B(t)u^*_t\, dt + s(t)'C(t)dw_t$$

$$d|y_t| = (\operatorname{sgn} y_t)[s'Bu^*_t dt + s'Cdw_t],$$

assuming that y spends negligible time at zero. Hence to drive $|y|$ to zero, we should set

$$u^*(t,x) = -\mathrm{sgn}[x\cdot s(t)]\ \mathrm{sgn}[B(t)'s(t)].$$

We shall verify that this u^* is extremal.

According to (9.12) we must show that a.e.(t)

$$\mathrm{sgn}\ E\{k_y(y_T)|\underline{F}_t^x\} = \mathrm{sgn}\ y_t,$$

where $y_t = x_t\cdot s(t)$ and

(9.13) $$dy_t = -(s'B)\mathrm{sgn}(B's)\ \mathrm{sgn}\ y_t dt + s'Cdw_t.$$

But now the Markov property implies that we need to show, a.e.(t)

(9.14) $$\mathrm{sgn}\ E\{k_y(y_T)|y_t\} = \mathrm{sgn}\ y_t.$$

The following argument which relies on the symmetry of Brownian motion does this. Let

$$\tau = \inf\{r \geq t: y_r = 0\}$$

$$z_r = 1_\tau(r)y_r + \big(1-1_\tau(r)\big)(-y_r) = 2[1_\tau(r)y_r] - y_r$$

$$\bar{w}_r = 1_\tau(r)w_r + \big(1-1_\tau(r)\big)\big(w_\tau-(w_r-w_\tau)\big) = 2[1_\tau(r)w_r]-w_r+2\big(1-1_\tau(r)\big)w_\tau$$

Then $\{\bar{w}_r\}$ is again a Brownian motion (cf. exercise 9.5.5), and since $\{z_r\}$ also satisfies (9.13), and the solutions of (9.13) are unique in law, cf. remark 2.5, then the distribution of z is the same as that of y. Hence we obtain

$$\begin{aligned} E\{k_y(y_T)|y_t\} &= E\{k_y(z_T)|y_t\} \\ &= E\{1_\tau(T)k_y(y_T) + \big(1-1_\tau(T)\big)k_y(-y_T)|y_t\} \\ &= E\{1_\tau(T)k_y(y_T)|y_t\} - E\{\big(1-1_\tau(T)\big)k_y(y_T)|y_t\} \end{aligned}$$

since k_y is odd. But since the left side of this equation is also the <u>sum</u> of the last two expectations then $E\{k_y(y_T)|y_t\} = E\{1_\tau(T)k_y(y_T)|y_t\}$, and for $T < \tau$,

$$\mathrm{sgn}\ k_y(y_T) = \mathrm{sgn}\ y_T = \mathrm{sgn}\ y_t \qquad \text{a.s.}$$

since k_y is odd and non-negative on the positive half-axis. This establishes (9.14), and hence u^* as given by (9.12) is extremal.

In the preceding problem we were essentially trying to minimize the distance from x_T to the plane $x\cdot v = 0$. Suppose we now add the constraint

$$E(v\cdot x_T) \leq \beta,$$

that is we want Ex_T to lie on one side of the plane $x \cdot v = \beta$. Let us also set $k(y) = y^2/2$. Thus the problem is

$$\inf\{E\ 1/2(v \cdot x_T)^2 : Ev \cdot x_T \le \beta\}$$

with the same state equation, i.e. (9.5), and the same set of controls $\mathcal{U}$. The Hamiltonian H does not change, but

$$p_t' = \lambda_0\ E\{v \cdot x_T | \underline{\underline{F}}_t^x\}\ v'\Phi(T,t) + \lambda_1 v'\Phi(T,t)$$

and u^* is extremal if

$$u^*(t,x) = \mathrm{sgn}[B(t)'p_t]$$

(9.15)

$$= \mathrm{sgn}\ E\{\lambda_0(v \cdot x_T) + \lambda_1 | \underline{\underline{F}}_t^x\}\mathrm{sgn}[B(t)'s(t)].$$

Let us set

$$R(\infty) = E\ v \cdot \{\Phi(T,0)x_0 - \int_0^T \Phi(T,t)B(t)\mathrm{sgn}[B(t)'s(t)]dt\}$$

$$R(\rho) = E\ v \cdot \{\Phi(T,0)x_0 - \int_0^T \Phi(T,t)B(t)\mathrm{sgn}[B(t)'s(t)]\mathrm{sgn}[x_t \cdot s(t)+\rho]dt\}$$

$$= v'\Phi(T,0)x_0 - \int_0^T |B(t)'s(t)|_1 E\ \mathrm{sgn}[x_t \cdot s(t) + \rho]dt$$

where $|B(t)'s(t)|_1 = s(t)'B(t)\mathrm{sgn}[B(t)'s(t)] = \sum_i |(B(t)'s(t))_i|$.

We begin by looking at the abnormal case, $\lambda_0 = 0$. Then we may take $\lambda_1 = -1$ and so u^* is extremal if

$$u^*(t,x) = -\mathrm{sgn}[B(t)'s(t)]$$

$$Ev \cdot x_T = \beta$$

(since we require $\lambda_1 J_1(u^*) = 0$). But this last equality means that $\beta = R(\infty)$, so only in this case do abnormal extremals arise, and moreover u^* is independent of k - all the control effort is concentrated on satisfying the constraint.

Let us now turn to the normal case. We set $\lambda = (-1,-\rho)$, $\rho \ge 0$. Comparing (9.15) and (9.12) leads us to guess

(9.16) $$u^*(t,x) = -\mathrm{sgn}[x \cdot s(t)+\rho]\mathrm{sgn}[B(t)'s(t)].$$

If we now set $z_t^\rho = x_t \cdot s(t)+\rho$ (with the above u^* determining x) then z^ρ satisfies (9.13), and we find just as before that

$$\mathrm{sgn}\ E\{z_T^\rho | z_t^\rho\} = \mathrm{sgn}\ z_t^\rho$$

and hence u^* as given by (9.16) is extremal, provided ρ is chosen such that $\lambda_1 J_1(u^*) = 0$ and $J_1(u^*) \leq 0$. The following observation allows us to characterize all cases. If we define

$$\tau_0 = \inf\{t \geq 0: z_t^\rho = 0\}$$

then $\operatorname{sgn} z_t^\rho = \operatorname{sgn} z_0^\rho$ for $t < \tau_0$, so that

$$E \operatorname{sgn} z_t^\rho = \operatorname{sgn} z_0^\rho \, E 1_{\tau_0}(t)$$

and

$$R(\rho) = v'\Phi(T,0)x_0 - \int_0^T |B(t)'s(t)|_1 \operatorname{sgn} z_0^\rho \, E\{1_{\tau_0}(t)\}dt$$

$$\downarrow R(\infty)$$

as $\rho \to \infty$ since as $\rho \to \infty$, $\tau_0 \uparrow T$, cf. exercise 9.5.6, and $\operatorname{sgn}[x_0 \cdot s(0)+\rho] \to 1$. Finally note that $Ev \cdot x_T = R(\rho)$. Thus

(i) if $\beta \geq R(0)$ then $\lambda = (-1,0)$ and an extremal control is

$$u^*(t,x) = -\operatorname{sgn}[x_t \cdot s(t)]\operatorname{sgn}[B(t)'s(t)],$$

(ii) if $R(0) > \beta > R(\infty)$ then there exists ρ^* such that $\beta = R(\rho^*)$, and if we take $\lambda = (-1,-\rho^*)$, then u^* given by (9.16) with $\rho = \rho^*$ is extremal,

(iii) if $\beta = R(\infty)$ then as we saw above we have the abnormal case $\lambda = (0,-1)$ and

$$u^*(t,x) = -\operatorname{sgn}[B(t)'s(t)],$$

(iv) if $\beta < R(\infty)$ then no admissible control exists, since for any control u

$$\begin{aligned} Ev \cdot x_T &= v'\Phi(T,0)x_0 + \int_0^T s(t)'B(t)Eu_t dt \\ &\geq v'\Phi(T,0)x_0 - \int_0^T s(t)'B(t)\operatorname{sgn}[B(t)s(t)']dt \\ &= R(\infty) > \beta. \end{aligned}$$

One can check that in the cases (i) and (ii) the constraint qualifications are met, i.e. there exists u in $\mathcal{U}_G$ such that $E^u x_T \cdot v < \beta$. Thus we might hope that in the normal case, the extremal control is optimal, cf. §11.

9.4 A non-linear example. We wish to find an extremal control for

$$\inf\{J(u): u \text{ in } \mathcal{U}\}$$

where

$$J(u) = E\{\int_0^T \ell(t,x_t)dt + c(x_T)\}$$

(9.17) $$dx_t = [F(t,x_t) + G(t,x_t)u_t]dt + C(t)dw_t$$

$$\mathcal{U} = \{u:[0,T]\times C^1 \to [-1,1],\text{ measurable, } \{\underline{\underline{G}}_t^1\} \text{ adapted such that (9.17) has a weak solution}\}$$

Here x_t lies in **R** i.e. $n = 1$, $x \to c(x)$ and $x \to \ell(t,x)$ are even, continuously differentiable, and increasing on $x > 0$,

$$|c(x)| + |\ell(t,x)| + |c_x(x)| + |\ell_x(t,x)| \le k_2(1+|x|^q),$$

$x \to F(t,x)$ is odd, $x \to G(t,x)$ is either even or odd, and both these functions are continuously differentiable, with bounded derivatives (uniformly in t), and finally

$$k_0^{-1} \le |C(t)| \le k_0.$$

The form of c and ℓ suggests that we minimize $|x_t|$, all t, and so we should take $xGu < 0$, i.e.

(9.18) $$u^*(t,x) = -\text{sgn}[x_t G(t,x_t)].$$

To verify that this candidate is extremal, observe that $H(t,x,u,p,\lambda) = p[F(t,x) + G(t,x)u] + \lambda\ell(t,x)$, $\lambda = -1$, and

$$p_t = -E\{c_x(x_T)\exp\int_t^T[F_x(s,x_s)+G_x(s,x_s)u_s^*]ds$$
$$+ \int_t^T \ell_x(s,x_s)\exp(\int_s^T[F_x(r,x_r)+G_x(r,x_r)u_r^*]dr)ds|\underline{\underline{F}}_t^x\}$$

Then u^* is extremal if

$$u^*(t,x) = \text{sgn } G(t,x_t)\text{sgn } p_t,$$

that is if

$$-\text{sgn } x_t = \text{sgn } p_t$$

or if

(9.19) $$\text{sgn } x_t = E\{c_x(x_T)\exp\int_t^T F_x(s,x_s) - G_x(s,x_s)\text{sgn}[x_s G(s,x_s)]ds$$
$$+ \ell_x(s,x_s)\exp\left(\int_s^T F_x(r,x_r) - G_x(r,x_r)\text{sgn}[x_r G(r,x_r)]dr\right)ds|x_t\}.$$

Since (9.18) gives

$$dx_s = [F(s,x_s) - |G(s,x_s)|\text{sgn } x_s]ds + C(s)dw_s$$

then again we can conclude that x_s and $-x_s$ have the same distribution given

$$s \geq \tau = \inf\{s \geq t,\ x_s = 0\}.$$

Since c_x, ℓ_x are odd, and $F_x - G_x \operatorname{sgn}[xG]$ is even then the right side of (9.19) equals

$$E\{1_\tau(T)\Big[c_x(x_T)\exp\int_t^T F_x - G_x \operatorname{sgn}[xG]\,ds$$

$$+ \int_t^T \ell_x(\exp\int_s^T F_x - G_x \operatorname{sgn}[xG]\,dr)\,ds\Big]\,|\,x_t\}.$$

Since for $T < \tau$, $\operatorname{sgn} x_s = \operatorname{sgn} x_t$ then a.s. the term inside the expectation is zero or has the same sign as x_t, hence (9.19) follows. Thus u^* is extremal.

9.5 Exercises

9.5.1 For the linear regulator with the constraint

$$E|x_T|^2 \leq \beta.$$

assume that x_0 is such that $E|x_T|^2 < \beta$ when the control $u \equiv 0$ is used. Show that no abnormal extremals can exist.

Hint: show that $E^u|x_T|^2 \leq E^0|x_T|^2$.

9.5.2 Find an extremal control for the linear regulator with constraint

$$E\ v'x_T = \beta.$$

9.5.3 Prove (9.6)

9.5.4 Prove (9.7)

9.5.5 Prove that $\{\bar{w}_r\}$ in example 9.3 is a Brownian motion on $(\Omega, \underline{\underline{F}}, \{\underline{\underline{F}}_t^x\}, P)$ if $\{w_r\}$ is.

9.5.6 Prove that $\tau_0 \uparrow T$ a.s. in example 9.3.

9.5.7 In example 9.3 show that the constraint qualifications are met in cases (i) and (ii).

9.5.8 If

(*) $\quad dx_t = b\ \beta(t)\cdot u_t\,dt + C(t)\,dw_t$

with b in $\mathbf{R}^n$, $\beta(t)$ in $\mathbf{R}^m$ bounded and measurable, C in $\mathbf{R}^n \otimes \mathbf{R}^n$ bounded, measurable with bounded inverse, let

$$J_0(u) = Ex_T'Dx_T,\quad J_{-1}(u) = Ex_T\cdot v - \alpha,$$

with $D \geq 0$ and $b \cdot v \neq 0$. The control problem is

$$\inf\{J_0(u): u \text{ in } \mathcal{U},\ J_{-1}(u) = 0\}$$

with $\mathcal{U} = \{u:[0,T]\times C^n \to [-1,1]^m$, measurable, $\{\underline{\underline{G}}_t^n\}$ adapted such that (*) has a weak solution$\}$. Show that exactly three cases arise:

(i) no admissible control exists if

$$|\alpha - v \cdot x_0| > |v \cdot b| \int_0^T \sum_i |\beta_i(t)| dt \equiv R(\infty),$$

(ii) if $|\alpha - v \cdot x_0| = R(\infty)$, then an abnormal extremal control is given by

$$u^*(t,x) = \operatorname{sgn}(\lambda_{-1} v \cdot b) \operatorname{sgn} \beta(t)$$

$$\lambda_{-1} = (\alpha - v \cdot x_0)/R(\infty),$$

(iii) if $|\alpha - v \cdot x_0| < R(\infty)$, then an extremal control is

$$u^*(t,x) = -\operatorname{sgn}[b'Dx + \rho^* b \cdot v]$$

where ρ^* is the solution of

$$\alpha - v \cdot x_0 = -(v \cdot b) \int_0^T E\, 1_\tau(t) \sum_i |\beta_i(t)| dt \operatorname{sgn}[b'Dx_0 + \rho b \cdot v]$$

where $\tau = \inf\{s \geq 0: b'Dx_s + \rho b \cdot v = 0\}$.

9.6 <u>Comments</u>. The linear regulator problem was solved long ago by using dynamic programming, cf. the book by Fleming and Rishel (1975). The predicted miss problem was first solved by Beneš (1976) also using dynamic programming in addition to a complicated approximation argument. Extremal controls for all three examples were found by Haussmann (1981a), see also Haussmann (1984).

10 Extremal controls and optimality

We know from section eight that optimal controls are extremal, and we have been able to find extremal controls for some very simple examples. We now show that in certain cases extremal controls are optimal. Let us begin by formulating the problem. The state x^u satisfies

$$(10.1) \qquad dx_t = f(t,x_t,u_t)dt + \sum_{k=1}^{d} [D^k(t)x_t + e^k(t)]dw_t^k,$$

on some fixed $(\Omega,\underline{\underline{F}},\{\underline{\underline{F}}_t\},P,\{w_t\})$. Here $\{w_t\}$ is a standard Brownian motion in $\mathbf{R}^d$, $f:[0,T]\times\mathbf{R}^n\times U \to \mathbf{R}^n$, $D^k: [0,T] \to \mathbf{R}^n\otimes\mathbf{R}^n$, $e^k:[0,T] \to \mathbf{R}^n$ for $k = 1,\cdots,d$ are all Borel measurable. We assume

(10.2) f is differentiable in x with

$$|f(t,x,u)| \le k_1(1+|x|+|u|)$$

$$|f_x(t,x,u)| + |D^k(t)| + |e^k(t)| \le k_1,$$

so that (10.1) has a unique solution. As we do not use Girsanov's theorem we do not require any invertibility of the diffusion coefficient.

As before we have

$$J_i(u) = E\{c_i(x_T) + \int_0^T \ell_i(t,x_t,u_t)dt\} \quad i = -m_1,\cdots,m_2.$$

We assume

(10.3) $c: \mathbf{R}^n \to \mathbf{R}^{m_1+1+m_2}$, $\quad \ell:[0,T]\times\mathbf{R}^n\times U \to \mathbf{R}^{m_1+1+m_2}$

are Borel measurable, differentiable in x and

$$|c(x)| + |\ell(t,x,u)| \le k_2(1+|x|^q+|u|^q)$$

$$|c_x(x)| + |\ell_x(t,x,u)| \le k_2(1+|x|^{q-1}+|u|^q)$$

for some q in $[1,\infty)$.

Then J(u) is well defined for any u which is a measurable, adapted, U-valued process such that $\int_0^T E|u_t|^{\bar{q}}dt < \infty$ if we assume

(10.4) $\qquad x_0$ is $\underline{\underline{F}}_0$ measurable and $E|x_0|^{\bar{q}} < \infty$.

As usual $\bar{q} > \max\{2,q\}$. Let us denote this class of controls by $\tilde{\mathcal{U}}$. It is the same as $\tilde{U}$ in §7, and depends of course on $(\Omega, \underline{\underline{F}}, \{\underline{\underline{F}}_t\}, P, \{w_t\})$.

The problem is

$$(10.5) \qquad \inf\{J_0(u): u \text{ in } \tilde{U},\ J_i(u) = 0 \quad i < 0,\ J_i(u) \le 0 \quad i > 0\}.$$

For this problem we define extremal controls as follows.

10.1 Definition. u^* in $\tilde{U}$ with corresponding solution x^* of (10.1) is extremal if there exists λ in $\mathbf{R}^{m_1+1+m_2}$, $\lambda \neq 0$, $\lambda_i \le 0$ if $i \ge 0$, $\lambda_i J_i(u^*) = 0$ if $i > 0$, such that for almost all t

$$\max_{u \in U} H(t, x^*_t, u, p_t, \lambda) = H(t, x^*_t, u^*_t, p_t, \lambda) \quad \text{a.s.}$$

where $p_t = E\{\bar{p}_t | \underline{\underline{F}}_t\}$ and

$$\bar{p}'_t = \lambda'\{c_x(x^*_T)\Phi(T,t) + \int_t^T \ell_x(s, x^*_s, u^*_s)\Phi(s,t)ds\}$$

$$(10.6) \qquad d\Phi(s,t) = f_x(s, x^*_s, u^*_s)\Phi(s,t)ds + \sum_{k=1}^{d} D^k(s)\Phi(s,t)dw^k_s, \quad \Phi(t,t) = I.$$

10.2 Remark. As noted in section 7, if u^* is extremal in the sense of definition 9.1, then u^* gives rise to a space $(\Omega, \underline{\underline{F}}, \{\underline{\underline{F}}_t\}, P, \{w_t\})$ and relative to this space u^* can be taken as an element of $\tilde{U}$, hence is extremal according to definition 10.1.

The following function H* corresponds to the Hamiltonian in the calculus of variations.

$$H^*(t,x,p,\lambda) = \sup_{u \in U} H(t,x,u,p,\lambda)$$

Observe that a.e.(t), $H^*(t, x^*_t, p_t, \lambda) = H(t, x^*_t, u^*_t, p_t, \lambda)$ a.s. if u^* is extremal.

10.3 Theorem. Assume (10.2) - (10.4) and assume that u^* is extremal, c_i is convex for $i \ge 0$, c_i is affine for $i < 0$, and $x \to H^*(t,x,p_t,\lambda)$ is concave a.e. (dtdP). Then u^* solves (10.5) if the problem is normal.

Proof: Let us write Φ_t for $\Phi(t,0)$, and let Ψ_t be the unique solution of

$$d\Psi_t = -\Psi_t[f_x(t, x^*_t, u^*_t) - \sum_k D^k(t)D^k(t)]dt - \Psi_t \sum_k D^k(t)dw^k_t$$

with $\Psi_0 = I$. Then $\Psi_t = \Phi_t^{-1}$ and $\Phi(s,t) = \Phi_s\Psi_t$ by the (pathwise) uniqueness of the solution of (10.6). Thus

$$\bar{p}'_t = \lambda'\{c_x(x^*_T)\Phi_T + \int_t^T \ell_x(s,x^*_s,u^*_s)\Phi_s ds\}\Psi_t$$

$$= \lambda'\{c_x(x^*_T)\Phi_T + \int_0^T \ell_x(s,x^*_s,u^*_s)\Phi_s ds\}\Psi_t - \lambda'\int_0^t \ell_x(s,x^*_s,u^*_s)\Phi_s ds\Psi_t$$

$$= \bar{p}'_0\,\Psi_t - \int_0^t \lambda'\ell_x(s,x^*_s,u^*_s)\Phi_s ds\Psi_t .$$

But if x^u satisfies (10.1) and u is in $\tilde{U}$ then

$$d(\Psi_t x^u_t) = -\Psi_t[f_x(t,x^*_t,u^*_t)x^u_t - f(t,x^u_t,u_t) + \sum_k D^k(t)e^k(t)]dt$$

$$+ \Psi_t \sum_k e^k(t)dw^k_t,$$

and even though $\bar{p}_t$ is <u>not</u> adapted we can compute

$$\bar{p}'_T x^u_T - \bar{p}'_0 x^u_0 = \bar{p}'_0\int_0^T d(\Psi_t x^u_t) - \int_0^T \lambda'\ell_x(s,x^*_s,u^*_s)\Phi_s\Psi_s x^u_s ds$$

$$- \int_0^T \int_0^t \lambda'\ell_x(s,x^*_s,u^*_s)\Phi_s ds\, d(\Psi_t x^u_t)$$

$$= \bar{p}'_0 \int_0^T - \Psi_t[f_x(t,x^*_t,u^*_t)x^u_t - f(t,x^u_t,u_t) + \sum_k D^k(t)e^k(t)]dt$$

$$+ \bar{p}'_0 \int_0^T \Psi_t \sum_k e^k(t)dw^k_t - \int_0^T \lambda'\ell_x(s,x^*_s,u^*_s)x^u_s ds$$

$$+ \int_0^T \int_0^t \lambda'\ell_x(s,x^*_s,u^*_s)\Phi_s ds\Psi_t[f_x(t,x^*_t,u^*_t)x^u_t - f(t,x^u_t,u_t)$$

$$+ \sum_k D^k(t)e^k(t)]dt - \int_0^T \int_0^t \lambda'\ell_x(s,x^*_s,u^*_s)\Phi_s ds\Psi_t \sum_k e^k(t)dw^k_t$$

$$= \int_0^T -\{\bar{p}'_t[f_x(t,x^*_t,u^*_t)x^u_t - f(t,x^u_t,u_t) + \sum_k D^k(t)e^k(t)]$$

$$+ \lambda'\ell_x(t,x^*_t u^*_t)x^u_t\}dt + \bar{p}'_0 \int_0^T \Psi_t \sum_k e^k(t)dw^k_t$$

$$- \int_0^T[\int_0^t \lambda'\ell_x(s,x^*_s,u^*_s)\Phi_s ds]\Psi_t \sum_k e^k(t)dw^k_t,$$

so that

$$\bar{p}_T'(x_T^u - x_T^*) = (\bar{p}_T'x_T^u - \bar{p}_0'x_0^u) - (\bar{p}_T'x_T^* - \bar{p}_0'x_0^*)$$

$$(10.7) \qquad = \int_0^T - [\bar{p}_t' f_x(t,x_t^*,u_t^*)+\lambda'\ell_x(t,x_t^*,u_t^*)](x_t^u-x_t^*)$$

$$+ \bar{p}_t'[f(t,x_t^u,u_t) - f(t,x_t^*,u_t^*)]dt.$$

Since c_i is convex and $\lambda_i \leq 0$ if $i \geq 0$, and c_i is affine if $i < 0$, then $\lambda'c(x)$ is concave and we obtain

$$\lambda'[c(x_T^u)+\int_0^T\ell(t,x_t^u,u_t)dt] - \lambda'[c(x_T^*)+\int_0^T\ell(t,x_t^*,u_t^*)dt]$$

$$\leq \lambda'c_x(x_T^*)(x_T^u-x_T^*) + \int_0^T \lambda'[\ell(t,x_t^u,u_t)-\ell(t,x_t^*,u_t^*)]dt$$

$$= \bar{p}_T'(x_T^u-x_T^*) + \int_0^T \lambda'[\ell(t,x_t^u,u_t)-\ell(t,x_t^*,u_t^*)]dt$$

$$= \int_0^T\{- H_x(t,x_t^*,u_t^*,\bar{p}_t,\lambda)(x_t^u-x_t^*)$$

$$+ H(t,x_t^u,u_t,\bar{p}_t,\lambda) - H(t,x_t^*,u_t^*,\bar{p}_t,\lambda)\}dt$$

where (10.7) was used for the last equality. Now we take expectations in this inequality and use the fact that the right side is linear in $\bar{p}$ and that $p_t = E\{\bar{p}_t|\underline{\underline{F}}_t\}$, to obtain

$$\lambda'J(u) - \lambda'J(u^*) \leq \int_0^T E\, E\{[- H_x(t,x_t^*,u_t^*,\bar{p}_t,\lambda)(x_t^u-x_t^*)$$

$$+ H(t,x_t^u,u_t,\bar{p}_t,\lambda) - H(t,x_t^*,u_t^*,\bar{p}_t,\lambda)]|\underline{\underline{F}}_t\}dt$$

$$= E \int_0^T[- H_x(t,x_t^*,u_t^*,p_t,\lambda)(x_t^u-x_t^*)$$

$$(10.8) \qquad + H(t,x_t^u,u_t,p_t,\lambda) - H(t,x_t^*,u_t^*,p_t,\lambda)]dt$$

$$\leq E \int_0^T[- H^*(t,x_t^u,p_t,\lambda) + H^*(t,x_t^*,p_t,\lambda)$$

$$+ H(t,x_t^u,u_t,p_t,\lambda) - H(t,x_t^*,u_t^*,p_t,\lambda)]dt$$

$$\leq 0$$

where the second inequality follows since $-H^*$ is convex and $-H_x(t,x_t^*,u_t^*,p_t,\lambda)$ is a subgradient of $-H^*(t,\cdot,p_t,\lambda)$ at x_t^*, by lemma 10.5.1. The last inequality follows from the definition of H^*.

Now if u satisfies the constraints of the problem (10.5) then

$$\lambda_0 J_0(u) \leq \lambda' J(u) \leq \lambda' J(u^*) = \lambda_0 J_0(u^*)$$

since for $i > 0$ $\lambda_i J_i(u) \geq 0$, $\lambda_i J_i(u^*) = 0$, and for $i < 0$, $J_i(u) = 0$ and $J_i(u^*) = 0$. Since the problem is normal (i.e. $\lambda_0 \neq 0$) then $\lambda_0 < 0$, so

$$J_0(u) \geq J_0(u^*)$$

and the result is established. Q.E.D.

10.4 Remarks

10.4.1. If the constraint qualifications are met relative to $\tilde{\mathcal{U}}$ rather than $\mathcal{U}_G$ then just as in the proof of theorem 6.4 we may conclude that no abnormal extremals exist.

10.4.2 Uniqueness of the solution of (10.1) implies that $\inf\{-\lambda' J(u): u \text{ in } \tilde{\mathcal{U}}\}$ is independent of $(\Omega, \underline{\underline{F}}, \{\underline{\underline{F}}_t\}, P, \{w_t\})$, and since any u in $\mathcal{U}_0$ can be embedded in some $\tilde{\mathcal{U}}$ such that

$$-\lambda' J(u) \geq \inf\{-\lambda' J(u): u \text{ in } \tilde{\mathcal{U}}\},$$

then

(10.9)
$$\inf_{\mathcal{U}_0}\{-\lambda' J(u)\} \geq \inf_{\tilde{\mathcal{U}}}\{-\lambda' J(u)\}.$$

Now if U is closed then theorem 7.4 gives equality in (10.9). But this now implies that if u^* is a normal extremal and if u is in $\mathcal{U}_0$ then

$$\lambda' J(u) \leq -\inf_{\mathcal{U}_0} - \lambda' J(u) = -\inf_{\tilde{\mathcal{U}}} - \lambda' J(u) \leq \lambda' J(u^*)$$

because of (10.8), so u^* solves

$$\inf\{J_0(u): u \text{ in } \mathcal{U}_0,\ J_i(u) = 0 \quad i < 0,\ J_i(u) \leq 0 \quad i > 0\}.$$

10.4.3 In general concavity a.e. of H* is difficult to establish, and is even false. However if

(10.10)
$$f(t,x,u) = A(t)x + b(t,u)$$

(10.11)
$$\ell(t,x,u) = \tilde{\ell}(t,x) + \bar{\ell}(t,u)$$

where a.e.(t) $x \to \tilde{\ell}_i(t,x)$ is convex for $i \geq 0$ and is affine for $i < 0$, then

$$H^*(t,x,p,\lambda) = p' A(t)x + \lambda' \tilde{\ell}(t,x) + h(t,p,\lambda)$$

where

$$h(t,p,\lambda) = \sup_{u\in U} \{p'b(t,u)+\lambda'\ell(t,u)\}$$

so that H* is concave in x, a.e.(t), for all (p,λ) such that $\lambda_i \leq 0$ if $i \geq 0$.

10.4.4 These results imply the optimality of the normal extremal controls found in examples 9.2 and 9.3 provided $k(\cdot)$ is convex. The optimality of u^* in example 9.4 will be settled in the next section.

10.5 Appendix. We set $H(t,x,u_t^*,p_t,\lambda) = \psi(x)$, $H^*(t,x,p_t,\lambda) = \phi(x)$.

10.5.1 Lemma. $\phi(x) - \phi(x_t^*) \leq \psi_x(x_t^*)(x-x_t^*)$

Proof: Since $(-\phi)$ is convex then a subgradient of $(-\phi)$ at x_t^* exists, i.e. there is a ν in $\mathbf{R}^n$ such that

$$\phi(x) - \phi(x_t^*) \leq \nu\cdot(x-x_t^*) \qquad \forall x \text{ in } \mathbf{R}^n. \tag{10.12}$$

Since $\psi(x) \leq \phi(x)$ with equality at $x = x_t^*$, then

$$\psi\big(\theta x+(1-\theta)x_t^*\big) - \psi(x_t^*) \leq \phi\big(\theta x+(1-\theta)x_t^*\big) - \phi(x_t^*)$$
$$\leq \theta\nu\cdot(x-x_t^*).$$

Dividing by θ and letting $\theta \downarrow 0$ and $\theta \uparrow 0$ shows that

$$\nu\cdot(x-x_t^*) = \frac{d}{d\theta}\,\psi\big(\theta x+(1-\theta)x_t^*\big)\Big|_{\theta=0} = \psi_x(x_t^*)(x-x_t^*).$$

The result follows from (10.12).

10.6 Exercises

10.6.1 Prove (10.12). Hint: let $A = \{(x,t):t \leq \phi(x)\}$ in theorem 4.3.

10.6.2 If Φ_t, Ψ_t are defined as in the proof of theorem 10.3, show that $\Psi_t = \Phi_t^{-1}$ and $\Phi(s,t) = \Phi_s\Psi_t$.

10.6.3 Prove the validity of remark 10.4.4.

10.6.4 Prove that the normal extremal found in exercise 9.5.8 is optimal.

10.7 Comments. The results of this section are contained in Haussmann (1982) and have been extended to the problem with partial observation in Haussmann (1982a). Similar results for the deterministic control problem can be found in Seierstad and Sydsaeter (1977).

11 Strongly extremal controls

We present now another method which can sometimes be used to establish optimality of an extremal control. Rather than produce a general theorem we shall content ourselves with applying the method to show that the extremal control found in example 9.4 is optimal. The idea is simple: show that the extremal control is a strong extremal to conclude by theorem 6.4 that it is optimal in $\mathcal{U}_G$. The difficulty lies in showing that it is a strong extremal, that is that it maximizes the Hamiltonian when the strong adjoint process, $\tilde{p}_t$, is used in place of p_t. Since

$$\tilde{p}_t' = \lambda' \chi_t \sigma(t,x_t)^+$$

then the first step is to obtain a representation of χ_t. We do this in some generality.

For the moment we assume that there is no control present. We are given a probability space $(\Omega, \underline{\underline{F}}, \{\underline{\underline{F}}_t\}, P)$ carrying $\{w_t\}$, $\{x_t\}$ satisfying

(11.1) $$dx_t = f(t,x_t)dt + \sigma(t,x_t)dw_t,$$

and we assume:

(11.2)
$$|f(t,x)|^2 \leq k_1(1+|x|^2), \quad |f_x(t,x)| \leq k_1;$$
$$|\sigma(t,x)| + |\sigma_x(t,x)| \leq k_1;$$
$$|\ell(t,x)| + |\ell_x(t,x)| + |c(x)| + |c_x(x)| \leq k_2(1+|x|^q);$$
$$E\exp(\varepsilon|x_0|^2) < \infty.$$

For x in C^n we write

$$L(x) = \int_0^T \ell(t,x_t)dt + c(x_T),$$

and as in (3.14) there exist $\{\chi_t\}$, $\{M_t\}$ such that

$$L(x) = L_0 + \int_0^T \chi_t dw_t + M_T \qquad \text{a.s.}$$

with $L_0 = E\{L(x)|\bar{\underline{\underline{F}}}_{0+}\}$, and such that $\{M_t\}$ is a square integrable martingale orthogonal to $\{w_t\}$ with $M_0 = 0$.

11.1 Theorem. Assume (11.2). Then

$$\chi_t = E\{\int_t^T \ell_x(s,x_s)\Phi(s,t)ds + c_x(x_T)\Phi(T,t)|\underline{\underline{F}}_t\}\sigma(t,x_t)$$

where

$$d\Phi(s,t) = f_x(s,x_s)\Phi(s,t)ds + \sum_k \sigma_x^k(s,x_s)\Phi(s,t)dw_s^k, \quad \Phi(t,t) = I.$$

Proof: Let $\{h_t : 0\leq t\leq T\}$ be any bounded, $\{\bar{\underline{\underline{F}}}_{t+}\}$ predictable $\mathbf{R}^d$-valued process, and let $\{x_t^\varepsilon : 0\leq t\leq T\}$ be the unique strong solution of

$$\begin{aligned} dx_t^\varepsilon &= f(t,x_t^\varepsilon)dt + \sigma(t,x_t^\varepsilon)[dw_t+\varepsilon h_t dt] \\ &= f(t,x_t^\varepsilon)dt + \sigma(t,x_t^\varepsilon)dw_t^\varepsilon \end{aligned}$$

where $\{w_t^\varepsilon\}$ is a standard Brownian motion on $(\Omega,\underline{\underline{F}},\{\bar{\underline{\underline{F}}}_{t+}\},P^\varepsilon)$ with $dP^\varepsilon = \zeta_T(-\varepsilon h)dP$. By strong uniqueness, hence law uniqueness, it follows that the law of $\{x_t\}$ on $(\Omega,\underline{\underline{F}},P)$ is the same as that of $\{x_t^\varepsilon\}$ on $(\Omega,\underline{\underline{F}},P^\varepsilon)$. Hence

$$E\,L(x) = E^\varepsilon L(x^\varepsilon)$$

and so

$$0 = \frac{d}{d\varepsilon} E^\varepsilon L(x^\varepsilon)\Big|_{\varepsilon=0} = \frac{d}{d\varepsilon} E\{\zeta_T(-\varepsilon h)L(x^\varepsilon)\}\Big|_{\varepsilon=0}. \tag{11.3}$$

But

$$\begin{aligned} \frac{d}{d\varepsilon}\zeta_T(-\varepsilon h)\Big|_{\varepsilon=0} &= \frac{d}{d\varepsilon}\exp\{-\varepsilon\int_0^T h_t\cdot dw_t - \frac{\varepsilon^2}{2}\int_0^T|h_t|^2dt\}\Big|_{\varepsilon=0} \\ &= \{-\int_0^T h_t\cdot dw_t - \varepsilon\int_0^T|h|^2dt\}\zeta_T(-\varepsilon h)\Big|_{\varepsilon=0} \\ &= -\int_0^T h_t\cdot dw_t, \end{aligned}$$

and, if L_x is the Fréchet derivative of L, then

$$\begin{aligned} \frac{d}{d\varepsilon} L(x^\varepsilon)_{\varepsilon=0} &= L_x(x^\varepsilon)y\Big|_{\varepsilon=0} \\ &= \int_0^T \ell_x(t,x_t)y_t dt + c_x(x_T)y_T \end{aligned}$$

where

$$y_t = \frac{d}{d\varepsilon} x_t^\varepsilon\Big|_{\varepsilon=0}.$$

The methods of section 8 allow us to compute $\{y_t\}$. Indeed if $\xi^\varepsilon = x^\varepsilon - x$, then

$$d\xi_t^\varepsilon = [f(t,x_t+\xi_t^\varepsilon)-f(t,x_t)]\,dt+\varepsilon\sigma(t,x_t+\xi_t^\varepsilon)h_t\,dt+[\sigma(t,x_t+\xi_t^\varepsilon)-\sigma(t,x_t)]\,dw_t$$

which is (8.3) with $F^\varepsilon(t,x) = \varepsilon\sigma(t,x)h_t$. With a minor change in the proof, lemma 8.1 holds, i.e.

$$E\|\xi^\varepsilon\|_T^p = O(\varepsilon^p).$$

Let y^ε be the solution of

$$dy_t^\varepsilon = f_x(t,x_t)y_t^\varepsilon dt + \varepsilon\sigma(t,x_t)h_t dt + \sum_k \sigma_x^k(t,x_t)y_t^\varepsilon dw_t^k,\ y_0^\varepsilon = 0,$$

then as in lemma 8.3

$$E\|\xi^\varepsilon - y^\varepsilon\|_T^p = o(\varepsilon^p) \quad \text{for } 2 \leq p < \bar{q},$$

and so

$$x_t^\varepsilon = x_t + y_t^\varepsilon + r_t^\varepsilon$$

where $E\|r^\varepsilon\|_T^p = o(\varepsilon^p)$. Hence $E\|y_t^\varepsilon - \varepsilon y_t\|_T^p = o(\varepsilon^p)$, i.e. $y_t = \varepsilon^{-1}y_t^\varepsilon|_{\varepsilon=0}$ and so

$$dy_t = f_x(t,x_t)y_t dt + \sigma(t,x_t)h_t dt + \sum_k \sigma_x^k(t,x_t)y_t dw_t^k,\ y_0 = 0,$$

i.e.

$$y_t = \int_0^t \Phi(t,s)\sigma(s,x_s)h_s ds.$$

We verify next that in (11.3) one may interchange $\frac{d}{d\varepsilon}$ and E.

$$\varepsilon^{-1}|\zeta_T(-\varepsilon h)L(x^\varepsilon) - L(x)| \leq \varepsilon^{-1}|\zeta_T(-\varepsilon h) - 1|\,|L(x^\varepsilon)| + \varepsilon^{-1}|L(x^\varepsilon) - L(x)|$$

and by Taylor's theorem and (11.2)

$$\varepsilon^{-1}|L(x^\varepsilon) - L(x)| = \left|\int_0^T \ell_x(t,x_t+\eta_t\xi_t^\varepsilon)\xi_t^\varepsilon dt + c_x(x_T+\eta_T\xi_T^\varepsilon)\xi_T^\varepsilon\right|$$

$$\leq K(1+\|x\|^q+\|\xi^\varepsilon\|^q)$$

where $0 \leq \eta_t(\omega) \leq 1$. Uniform integrability implies

$$\lim_{\varepsilon\to 0} E\{\frac{L(x^\varepsilon)-L(x)}{\varepsilon}\} = E \lim_{\varepsilon\to 0}\{\frac{L(x^\varepsilon)-L(x)}{\varepsilon}\}.$$

Moreover again by Taylor's theorem, for some η_0 in [0,1]

(11.4) $$\varepsilon^{-1}|\zeta_T(-\varepsilon h)-1|\,|L(x^\varepsilon)|$$

$$= |-\int_0^T h_t\cdot dw_t - \eta_0\varepsilon\int_0^T|h_t|^2dt|\ \zeta_T(-\varepsilon\eta_0 h)\ |L(x^\varepsilon)|$$

By theorem 2.2, for any $p > 1$, there exists k such that $E|\zeta_T(-\varepsilon\eta_0 h)|^p \leq k$ for all ε sufficiently small depending on p. The other factors on the right side of (11.4) are integrable to any power by (11.2), so again uniform integrability implies

$$\lim_{\varepsilon\to 0} E\{[\frac{\zeta_T(-\varepsilon h)-1}{\varepsilon}]L(x^\varepsilon)\} = E\ \lim_{\varepsilon\to 0}\{[\frac{\zeta_T(-\varepsilon h)-1}{\varepsilon}]L(x^\varepsilon)\}.$$

It now follows from (11.3) that

$$0 = E\{\frac{d}{d\varepsilon}\zeta_T(-\varepsilon h)L(x^\varepsilon)|_{\varepsilon=0}\}$$

$$= E\{-\int_0^T h_t\cdot dw_t L(x) + \int_0^T \ell_x(t,x_t)\int_0^t \Phi(t,s)\sigma(s,x_s)h_s dsdt$$

$$+ c_x(x_T)\int_0^T\Phi(T,s)\sigma(s,x_s)h_s ds\}$$

so that, cf. appendix 3.8,

$$E\int_0^T \chi_s h_s ds = E\{\int_0^T h_s\cdot dw_s[L_0+\int_0^T \chi_s dw_s + M_T]\}$$

$$= E\{\int_0^T[c_x(x_T)\Phi(T,s)+\int_s^T \ell_x(t,x_t)\Phi(t,s)dt]\sigma(s,x_s)h_s ds\}.$$

Since $\{h_s\}$ is an arbitrary predictable process and since $\{\chi_s\}$ is predictable then we may deduce that for almost all s

$$\chi_s = E\{c_x(x_T)\Phi(T,s) + \int_s^T \ell_x(t,x_t)\Phi(t,s)dt|\underline{F}_s\}\sigma(s,x_s) \quad \text{a.s.}$$

and the result follows. Q.E.D.

If we now allow f and ℓ to depend on u we obtain

11.2 Corollary. Assume (11.2) with the obvious replacements:

$$|f(t,x,u)|^2 \leq k_1(1+|x|^2+|u|^2),\ |f_x(t,x,u)| \leq k_1;$$

$$|\ell(t,x,u)| + |\ell_x(t,x,u)| + |c(x)| + |c_x(x)| \leq k_2(1+|x|^q+|u|^q);$$

and assume that $u_t = u(t,x_t)$ with

$$u(t,x) \leq k_u(1+|x|),\ |u_x(t,x)| \leq k_u.$$

Then

$$\int_0^T \ell(t,x_t,u_t)dt + c(x_T) = L_0 + \int_0^T \chi dw_t + M_T$$

with $L_0 = E\{\int_0^T \ell(t,x_t,u_t)dt + c(x_T)\}$, $\{M_t\}$ orthogonal to $\{w_t\}$, and

$$\chi_t = E\{\int_t^T [\ell_x(s,x_s,u(s,x_s)) + \ell_u(s,x_s,u(s,x_s))u_x(s,x_s)]\Phi(s,t)ds$$
$$+ c_x(x_T)\Phi(T,t)|\underline{\underline{F}}_t\}\sigma(t,x_t)$$
$$d\Phi(s,t) = [f_x(s,x_s,u(s,x_s)) + f_u(s,x_s,u(s,x_s))u_x(s,x_s)]\Phi(s,t)ds$$
$$+ \sum_k \sigma_x^k(s,x_s)\Phi(s,t)dw_s^k, \ \Phi(t,t) = I.$$

Hence if an extremal control is Markovian and smooth, then we can compute the putative strong adjoint process by corollary 11.2.

Let us now apply this result to show that the extremal control of example 9.4 is optimal. We recall that

$$dx_t = [F(t,x_t) + G(t,x_t)u_t]dt + C(t)dw_t$$
$$L(x) = \int_0^T \ell(t,x_t)dt + c(x_T)$$
$$\sigma(t,x)^+ = C(t)^{-1}$$

and that

$$u^*(t,x) = -\mathrm{sgn}[x_t G(t,x_t)]$$

is an extremal control, but to apply corollary 11.2 we require a smooth control. For $\varepsilon > 0$ let us set

$$u^\varepsilon(t,x) = v^\varepsilon(xG(t,x))$$
$$v^\varepsilon(y) = \begin{cases} -\mathrm{sgn}\ y & \text{if } |y| \geq \varepsilon \\ -\sin(\pi y/2\varepsilon) & \text{if } |y| \leq \varepsilon \end{cases}$$

Observe that $u^*(t,\cdot)$ and $u^\varepsilon(t,\cdot)$ are even or odd as $G(t,\cdot)$ is even or odd. Now we set

$$\theta_t^\varepsilon = C(t)^{-1}G(t,x_t)[u^\varepsilon(t,x_t)-u^*(t,x_t)]$$

so that

(11.5) $$dx_t = [F(t,x_t) + G(t,x_t)u^\varepsilon(t,x_t)]dt + C(t)dw_t^\varepsilon$$

where, by Girsanov's theorem, $\{w_t^\varepsilon\}$ is a Brownian motion on $(\Omega,\underline{\underline{F}},\{\underline{\underline{F}}_t^x\},P^\varepsilon)$, $dP^\varepsilon = \zeta_T(\theta^\varepsilon)dP$. Our regularity hypotheses also imply that (11.5) has a unique solution which is strong, so

$$F_{=t}^{x} = F_{=t} = F_{=t}^{w^{\varepsilon}}$$

and so in the martingale representation

$$L(x) = L_0 + \int_0^T \chi_t dw_t^{\varepsilon} + M_T^{\varepsilon}$$

we have $M_T^{\varepsilon} = 0$ a.s. Corollary 11.2 implies that

(11.6) $$-\chi_t^{\varepsilon} C(t)^{-1} = -E\{c_x(x_T)\Phi^{\varepsilon}(T,t) + \int_t^T \ell_x(s,x_s)\Phi^{\varepsilon}(s,t)ds \mid F_{=t}^{x}\}$$

(11.7) $$\Phi^{\varepsilon}(s,t)=\exp \int_t^s [F_x(r,x_r)+G_x(r,x_r)u^{\varepsilon}(r,x_r)+G(r,x_r)u_x^{\varepsilon}(r,x_r)]dr.$$

Now observe that Φ^{ε} is "even" in $\{x_t\}$, i.e. if $\{x_t\}$ is replaced by $\{-x_t\}$, then Φ^{ε} does not change. From (11.6) we can conclude just as in section 9, that

(11.8) $$\chi_t^{\varepsilon} C(t)^{-1} = 0 \quad \text{if } x_t = 0,$$
$$\operatorname{sgn}[\chi_t^{\varepsilon} C(t)^{-1}] = \operatorname{sgn} x_t \quad \text{otherwise,}$$

and so, for u in U, and for x_t such that $|x_t G(t,x_t)| \geq \varepsilon$, we have

(11.9) $$H(t,x_t,u,\tilde{p}_t^{\varepsilon},-1) \leq H(t,x_t,u_t^{\varepsilon},\tilde{p}_t^{\varepsilon},-1) \quad \text{a.s.}$$

with

$$\tilde{p}_t^{\varepsilon} = -\chi_t^{\varepsilon} C(t)^{-1}.$$

Note that a.s. can be taken with respect to either P or P^{ε}; they are equivalent. For x_t such that $x_t G(t,x_t) = 0$, (11.9) holds with equality, because

$$H(t,x,u,p,-1) = [pF(t,x) - \ell(t,x)] + pG(t,x)u$$

is independent of u if $pG(t,x) = 0$, which is the case when $xG(t,x) = 0$, cf. (11.8). Hence (11.9) will also hold in the limit as $\varepsilon \downarrow 0$, and this for <u>all</u> x_t.

Since we know that $u_t^{\varepsilon} \to u_t^{*}$, and since H is continuous in (u,p), then in the limit as $\varepsilon \downarrow 0$ (11.9) implies that u^{*} is a strong extremal (hence optimal by theorem 6.4), provided for almost all t $\tilde{p}_t^{\varepsilon} \to \tilde{p}_t$ a.s., the strong adjoint process.

It remains to show, then, that there exists $\{\chi_t\}$ such that $\chi_t^\varepsilon \to \chi_t$ and

(11.10) $$L(x) = L_0 + \int_0^T \chi_t dw_t .$$

But for each t $\theta_t^\varepsilon \to 0$ a.s. so that $\zeta_T(\theta^\varepsilon) \to 1$ in probability (P) and, by uniform integrability, in L^p for some $p > 1$, cf. corollary 2.3. But $L(x)$ is in $L^{p'}$ for any $p' < \infty$, so

$$E^\varepsilon L(x) = E\, \zeta_T(\theta^\varepsilon) L(x) \to E\, L(x).$$

Since $dw^\varepsilon = dw - \theta^\varepsilon dt$, then

(11.11) $$\int_0^T \chi^\varepsilon dw = L(x) - L_0(x) + \int_0^T \chi^\varepsilon \theta^\varepsilon dt.$$

Moreover as $\varepsilon \to 0$

$$E\{|\int_0^T \chi^\varepsilon \theta^\varepsilon dt|^2\}$$

$$\leq E\{(\int_0^T |\chi^\varepsilon|^2 dt)^2\}^{1/2}\, E\{(\int_0^T |\theta^\varepsilon|^2 dt)^2\}^{1/2}$$

$$\to 0$$

if $E\{(\int_0^T |\chi^\varepsilon|^2 dt)^2\}$ is bounded. But

$$E\{(\int_0^T |\chi^\varepsilon|^2 dt)^2\} = E^\varepsilon\, \zeta_T^\varepsilon(-\theta^\varepsilon)(\int_0^T |\chi^\varepsilon|^2 dt)^2$$

$$\leq E^\varepsilon\{|\zeta_T^\varepsilon(-\theta^\varepsilon)|^p\}^{1/p}\, E^\varepsilon\{(\int_0^T |\chi^\varepsilon|^2 dt)^{2p'}\}^{1/p'}$$

where

$$\zeta_T^\varepsilon(-\theta^\varepsilon) = \exp\{\int_0^T (-\theta_t^\varepsilon) dw_t^\varepsilon - \frac{1}{2}\int_0^T |-\theta_t^\varepsilon|^2 dt\}.$$

Since $\{x_t\}$ satisfies (11.5) and since

$$|F| + |Gu^\varepsilon| \leq |F| + |G| \leq k_1(1+|x|),\quad |C(t)| \leq k_0,$$

and

$$|-\theta_t^\varepsilon| \leq |C(t)^{-1}|\,|G(t,x_t)|\,|u^\varepsilon - u^*| \leq k_0 k_1 (1+|x_t|)$$

then by corollary 2.3 again $E^\varepsilon|\zeta_T^\varepsilon(-\theta^\varepsilon)|^p \leq k$ for some $p > 1$ and some k, both independent of ε. Finally as in (3.19)

$$E^{\varepsilon}(\int_0^T|\chi^{\varepsilon}|^2dt)^{2p'} = E^{\varepsilon}([L_{\bullet}-L_0]_T)^{2p'}$$
$$\leq \bar{c}_1 E^{\varepsilon}\{\|L_{\bullet}-L_0\|_T^{4p'}\}$$
$$\leq \bar{c}\ E^{\varepsilon}|L_T-L_0|^{4p'}$$
$$\leq \bar{c}\ E\{\zeta_T(\theta^{\varepsilon})^p\}^{1/p}\ E\{|L_T-L_0|^{4p'p'}\}^{1/p'}$$

which is bounded uniformly in ε by corollary 2.3. Hence by (11.11)

(11.12) $$\int_0^T \chi^{\varepsilon}dw \to L(x) - L_0 \quad \text{in } L^2(dP).$$

Let $\varepsilon = \frac{1}{n}$ and write χ^n for χ^{ε}. Since

$$E|\int_0^T \chi^n dw - \int_0^T \chi^m dw|^2 = E\int_0^T |\chi^n - \chi^m|^2 dt,$$

then $\{\chi^n\}$ is Cauchy in $L^2(dtdP)$, and hence by completeness there exists χ in $L^2(dtdP)$ such that $\chi^n \to \chi$. Now (11.12) implies that (11.10) is satisfied, and moreover since the χ^n are predictable, then so is χ. This establishes the optimality of u^*.

11.3 Remark. It is not at all clear that when we pass to the limit ($\varepsilon \to 0$) in (11.6), we obtain

$$-\chi_t C(t)^{-1} = -E\{c_x(x_T)\Phi(T,t) + \int_t^T \ell_x(s,x_s)\Phi(s,t)ds|\underline{\underline{F}}_t^x\}$$

i.e. $\tilde{p}_t = p_t$,

because $|u_x^{\varepsilon}| \to \infty$ as $\varepsilon \to 0$. In general the terms $f_u u_x, \ell_u u_x$ appear in the representation for χ, but for an optimal u^* they vanish. Intuitively this follows from optimality since (for smooth f,ℓ).

(11.13) $$\frac{\partial H}{\partial u}(t,x_t,u,p_t,\lambda)|_{u=u^*} = 0$$

and (for $\sigma = \sigma(t)$)

$$dp_t = -\frac{\partial H}{\partial x}(t,x_t,u_t^*,p_t,\lambda)dt$$

$$d\tilde{p}_t = -\frac{dH}{dx}(t,x_t,u(t,x_t^*),p_t,\lambda)dt$$
$$= -[\frac{\partial H}{\partial x}(t,x_t,u_t^*,p_t,\lambda) + \frac{\partial H}{\partial u}(t,x_t,u_t^*,p_t,\lambda)u_x^*(t,x_t)]dt$$
$$= dp_t$$

by (11.13).

In fact in Haussmann (1981) it is shown under reasonable hypotheses, that equality holds, that is

$$p_t = \tilde{p}_t = - V_x(t,x_t)$$

where V is the value function of the problem (6.7).

Here is another approach to proving optimality based on this characterization of the adjoint process. Since there are no constraints in example 9.5 then

$$J_0(u) = -\lambda \cdot J(u)$$

if λ is normalized so that $\lambda = \lambda_0 = -1$; hence after this normalization the value function of (6.7) is independent of λ, and hence the adjoint process (strong or not) is unique, i.e. is the <u>same</u> for all normal extremals. Provided that an optimal control <u>exists</u>, one can now argue that any extremal control for example 9.4 is given <u>uniquely</u> in terms of p_t by

$$u^*(t,x) = \mathrm{sgn}[G(t,x_t)p_t]$$

hence must be the optimal control.

11.4 Exercises

11.4.1 Show that

$$E\|\xi^\varepsilon\|_t^p = O(\varepsilon^p)$$

in the proof of theorem 11.1.

11.4.2 Prove (11.7).

11.4.3 Use (11.8) to prove (11.9).

11.4.4 Use corollary 11.2 to find a strongly extremal (hence optimal) control for the linear regulator.

11.4.5 Use the method applied to example 9.4 to show that the normal extremals found for example 9.3 are strong extremals and hence optimal.

11.5 <u>Comments</u>. The method used in this section to prove optimality has in common with Beneš (1975) the feature that the controls are regularized. However whereas Beneš could not prove the optimality of the control

$$u^*(t,x) = - \operatorname{sgn}[x_t G(t,x_t)],$$

this was not too difficult by our method.

The proof of theorem 11.1 given here follows ideas used by Bismut in his development of the Malliavin calculus. The theorem was originally proved in greater generality by Haussmann (1979).

12 Other necessary conditions

Let us now compare our results to some necessary conditions found in the literature. These are mostly concerned with controls which give rise to strong solutions. We begin by defining a few classes of controls. A fixed probability space carrying a Brownian motion $\{w_t: 0 \leq t \leq T\}$ is given. Let

$$\mathcal{U}_N^s = \{u: [0,T]\times\Omega \to U, \text{ measurable, } \{\underline{\underline{F}}_t^w\} \text{ adapted, } \int_0^T E|u|^q dt < \infty\}.$$

Assume that

$$\begin{aligned} &|f(t,x,u) - f(t,y,u)| + |\sigma(t,x) - \sigma(t,y)| \leq k|x-y|, \\ \textbf{(12.1)} \quad &|f(t,x,u)| + |\sigma(t,x)| \leq k(1+|x|+|u|). \end{aligned}$$

Then for u in $\mathcal{U}_N^s$

(12.2) $$dx_t = f(t,x_t,u_t)dt + \sigma(t,x_t)dw_t,$$

always has a unique strong solution whose q^{th} moment is finite if the same is true of x_0. Now let

$$\mathcal{U}_A^s = \{u \in \mathcal{U}_N^s: u \text{ is } \{\underline{\underline{F}}_t^x\} \text{ adapted}\},$$

and let $\mathcal{U}_0^s$ be the set of measurable, $\{\underline{\underline{G}}_t^n\}$ adapted functions $u: [0,T]\times C^n \to U$ for which (12.2) has a strong solution if $u_t(\omega) = u(t,x(\omega))$, such that

$$\int_0^T E|u(t,x(\omega))|^q dt < \infty.$$

Then

$$\mathcal{U}_0^s \hookrightarrow \mathcal{U}_A^s \subset \mathcal{U}_N^s$$

where $\hookrightarrow$ denotes a canonical imbedding. Observe that $\mathcal{U}_0^s$ is the only class of any physical interest, since underlying probability spaces are always figments of our imagination. On the other hand only $\mathcal{U}_N^s$ is explicitly defined, so that only this case is tractable, i.e. one establishes necessary conditions satisfied by u^*, a solution of

$$\inf\{J_0(u): u \in \mathcal{U}_N^s, J_i(u) = 0 \quad i < 0, J_j(u) \leq 0 \quad j > 0\}.$$

This is essentially done by Kushner (1972), by Arkin and Saksonov (1979), and in the absence of constraints by Bensoussan (1982a).

Moreover our theorem 8.9 gives the solution since $\mathcal{U}_N^s = \tilde{\mathcal{U}}$. Note that if we can exhibit a u^* in $\mathcal{U}_0^s$ which satisfies these necessary conditions (they are the same as ours) then we can consider it as a candidate for an optimal control. Unfortunately it is difficult in general to verify that a control u^* is in $\mathcal{U}_0^s$.

In the absence of constraints, if σ is invertible and σ, σ^{-1} are bounded, then we can say more. In this case for u in $\mathcal{U}_A^s$ $\underline{\underline{F}}_t^x = \underline{\underline{F}}_t^w$ (cf. Bensoussan 1982a) so that $\bar{\underline{\underline{G}}}_{t+}^n = \bar{\underline{\underline{G}}}_t^n$ because $\bar{\underline{\underline{F}}}_{t+}^w = \bar{\underline{\underline{F}}}_t^w$ (a property of Brownian motion, cf. exercise 0.3.5). Since any u in $\mathcal{U}_A^s$ can be written as v(t,x) for some $\{\underline{\underline{G}}_{t+}^n\}$ adapted v, i.e. v can be modified to be an element of $\mathcal{U}_0^s$, then

$$\mathcal{U}_A^s = \mathcal{U}_0^s.$$

Moreover Viot has shown (cf. Bensoussan 1982a) that $\mathcal{U}_A^s$ is dense in $\mathcal{U}_N^s$ in the L^2 topology, so that any solution of

(12.3) $$\inf\{J_0(u):\ u \in \mathcal{U}_0^s\}$$

also solves

(12.4) $$\inf\{J_0(u):\ u \in \mathcal{U}_N^s\}$$

and must then satisfy the necessary conditions in $\mathcal{U}_N^s$. This fact is comforting to know, but still does little to ascertain when a certain $u^*(t,x)$ is in $\mathcal{U}_0^s$. The following deep result due to Veretennikov (1981) is helpful here. Let

$$\mathcal{U}_M = \{u:\ [0,T]\times\mathbf{R}^n \to U,\ \text{Borel measurable},\ |u(t,x)| \le K_u(1+|x|)\}$$

Then $\mathcal{U}_M \subset \mathcal{U}_0$ so a weak solution exists, and the surprising result is that any weak solution is strong, thus

$$\mathcal{U}_M \subset \mathcal{U}_0^s.$$

Now if we find a u^* in $\mathcal{U}_M$ which satisfies the necessary conditions in $\mathcal{U}_N^s$, then it is a likely candidate for a solution of (12.3) or (12.4).

In connection with this result, one should ask in what class do optimal controls exist. Using dynamic programming Krylov (1980) has shown that the infima in (12.3) and (12.4) are equal without any non-degeneracy assumption, and if $\sigma\sigma^*$ is non-degenerate, then these infima are equal to

$$\inf\{J_0(u):\ u \in \mathcal{U}_M\},$$

which infimum is attained, so that an optimal Markovian control exists. The last part of the statement follows from a measurable selection theorem and Veretennikov's result. Hence if σ is non-degenerate and if no constraints are present, then the theory of necessary conditions with strong solutions is satisfactory.

Let us mention that Bismut (1973) has also derived necessary conditions using convex analysis, and allowing f and σ to depend on ω; however the convexity requirements seem to be quite stringent, and constraints are not included.

Let us now turn to the theory of weak solutions. Let $\mathcal{U}_N$ be the set of U-valued adapted stochastic processes u_t on some filtered probability space which may be different for different processes, such that $E\int_0^T |u_t|^q dt < \infty$. Again an adapted solution of (12.2) exists since we still assume (12.1). If $\mathcal{U}_A = \{u \in \mathcal{U}_N : u \text{ is } \{\underline{\underline{F}}_t^x\} \text{ adapted}\}$ then

(12.5) $$\mathcal{U}_M \hookrightarrow \mathcal{U}_0 \hookrightarrow \mathcal{U}_A \subset \mathcal{U}_N$$

and $\mathcal{U}_*^s \subset \mathcal{U}_*$ for $* = 0$, A, N. Note the $\mathcal{U}_M$, $\mathcal{U}_0$, $\mathcal{U}_N$ are all explicitly defined, but $\mathcal{U}_0$ is the preferred class to work with since $\mathcal{U}_M$ may be too small and $\mathcal{U}_N$ is certainly too large. Theorem 7.4 shows that in the absence of constraints

$$\inf\{J_0(u) : u \in \mathcal{U}_0\} = \inf\{J_0(u) : u \in \mathcal{U}_N^s\},$$

but the same proof shows in fact that

(12.6) $$\inf\{J_0(u) : u \in \mathcal{U}_0\} = \inf\{J_0(u) : u \in \mathcal{U}_N\}.$$

Under non-degeneracy assumptions Bismut (1976) and El Karoui (1981) have shown that

$$\inf\{J_0(u) : u \in \mathcal{U}_M\}$$

has a solution, which also solves

$$\inf\{J_0(u) : u \in \mathcal{U}_A\},$$

hence even

$$\inf\{J_0(u) : u \in \mathcal{U}_N\}$$

by (12.5) and (12.6). Haussmann (1985) shows this directly when non-degeneracy is replaced by a convexity hypothesis.

In summary, when applied to a Markovian u^* (the usual case) the results of Kushner and Bensoussan are equivalent to ours in the absence of constraints. But when constraints are present one can no longer justify using a class of controls larger than $\mathcal{U}_0^s$ or $\mathcal{U}_0$, so that our result would be the preferred one.

Finally we mention that these results based on strong variations require σ to be independent of u. If σ depends on u then one can use weak variations to obtain (weaker) necessary conditions. This is done by Bismut (1973) and Bensoussan (1982a) with $\mathcal{U} = \mathcal{U}_N^s$ and with no constraints. It is clear that one may add constraints without too much difficulty, but it is unlikely that one can take a more practical $\mathcal{U}$, e.g. $\mathcal{U}_0$.

A more complete survey can be found in Haussmann (1985a).

References

Arkin, V.I. and Saksonov, M.T. (1979), Necessary optimality conditions for stochastic differential equations, Soviet Math. Dokl., 20, 1-5.

Beneš, V.E. (1970), Existence of optimal strategies based on specified information for a class of stochastic decision problems, SIAM J. Control, 8, 171-180.

Beneš, V.E. (1971), Existence of optimal stochastic controls, SIAM J. Control, 9, 446-472.

Beneš, V.E. (1975), Composition and invariance methods for solving some stochastic control problems, Adv. Appl. Prob., 7, 299-329.

Beneš, V.E. (1976), Full "bang" to reduce predicted miss is optimal, SIAM J. Control Optim., 14, 62-84.

Bensoussan, A. (1982), Stochastic Control by Functional Analysis Methods, North-Holland, Amsterdam.

Bensoussan, A. (1982a), Lectures on stochastic control, Lecture Notes in Mathematics, 972, 1-62.

Bismut, J.M. (1973), Conjugate convex functions in optimal stochastic control, J. Math. Anal. Applic., 44, 384-404.

Bismut, J.M. (1976), Théorie probabiliste du contrôle des diffusions, Mem. Amer. Math. Soc. No. 176.

Davis, M.H.A. (1973), On the existence of optimal policies in stochastic control, SIAM J. Control, 11, 587-594.

Dellacherie, C. and Meyer, P.A. (1975), Probabilités et potentiel, chapitres I a IV, Hermann, Paris.

Dellacherie, C. and Meyer, P.A. (1980), Probabilités et potentiel, chapitres V a VIII, Hermann, Paris.

Doob, J.L. (1953), Stochastic Processes, Wiley, New York.

Dunford, N. and Schwartz, J.T. (1965), Linear Operators, Part I, Wiley, New York.

El Karoui, N. (1981), Les aspects probabiliste du contrôle stochastique, Lecture Notes in Mathematics, 876, 74-239.

Fleming, W.H. and Pardoux, E. (1982), Existence of optimal controls for partially observed diffusions, SIAM J. Control Optim., 20, 261-283.

Fleming, W.H. and Rishel, R.W. (1975), Deterministic and Stochastic Optimal Control, Springer, New York.

Girsanov, I.V. (1960), On transforming a certain class of stochastic processes by absolutely continuous substitution of measures, Theory Prob. Appl., 5, 285-301.

Halmos, P.R. (1950), Measure Theory, Van Nostrand Reinhold, New York.

Haussmann, U.G. (1976), General necessary conditions for optimal control of stochastic systems, Math. Programming Studies, 6, 30-48.

Haussmann, U.G. (1979), On the integral representation of functionals of Itô processes, Stochastics, 3, 17-27.

Haussmann, U.G. (1981), On the adjoint process for optimal control of diffusion processes, SIAM J. Control Optim., 19, 221-243.

Haussmann, U.G. (1981a), Some examples of optimal stochastic controls, SIAM Review, 23, 292-307.

Haussmann, U.G. (1982), Extremal controls for completely observable diffusions, Lecture Notes in Control and Information Sciences, 42, 149-160.

Haussmann, U.G. (1982a), Optimal control of partially observed diffusions via the separation principle, Lecture Notes in Control and Information Sciences, 43, 302-311.

Haussmann, U.G. (1984), Extremals in stochastic control theory, Lecture Notes in Control and Information Sciences, 59, 461-470.

Haussmann, U.G. (1985), Existence of optimal Markovian controls for degenerate diffusions, Lecture Notes in Control and Information Sciences, 78, 171-186.

Haussmann, U.G. (1985a), The maximum principle for the optimal control of stochastic differential equations, to appear in Ann. CEREMAD, Birkhäuser.

Ichikawa, A. (1978), Notes on a maximum principle of Haussmann, I.A.M.S. Tech. Rept. 78-11, U.B.C., Vancouver.

Ikeda, N. and Watanabe, S. (1981), Stochastic Differential Equations and Diffusion Processes, North-Holland, Amsterdam.

Krylov, N.V. (1980), Controlled Diffusion Processes, Springer, New York.

Kushner, H.J. (1965), On the stochastic maximum principle: Fixed time of control, J. Math. Anal. Applic., 11, 78-92.

Kushner, H.J. (1972), Necessary conditions for continuous parameter stochastic optimization problems, SIAM J. Control, 10, 550-565.

Liptser, R.S. and Shiryayev, A.N. (1977), Statistics of Random Processes I, General Theory, Springer, New York.

McKean, H.P. (1969), Stochastic Integrals, Academic Press, New York.

Neustadt, L. (1976), Optimization, Princeton U. Press, Princeton.

Novikov, A.A. (1972), On an identity for stochastic integrals, Theory Probability Applic., 17, 717-720.

Seierstad, A. and Sydsaeter K. (1977), Sufficient conditions in optimal control theory, Int. Economic Rev., 18, 367-391.

Veretennikov, A. Ju. (1981), On strong solutions and explicit formulas for solutions of stochastic integral equations, Math. USSR Sbornik, 39, 387-403.

Wong, E. (1971), Representation of martingales, quadratic variations and applications, SIAM J. Control, 9, 621-633.